Green Energy and Technology

Vasily Y. Ushakov

Electrical Power Engineering

Current State, Problems and Perspectives

 Springer

Vasily Y. Ushakov
Tomsk Polytechnik University
Tomsk
Russia

ISSN 1865-3529　　　　　　　ISSN 1865-3537　(electronic)
Green Energy and Technology
ISBN 978-3-319-87285-8　　　ISBN 978-3-319-62301-6　(eBook)
https://doi.org/10.1007/978-3-319-62301-6

This Springer imprint is published by Springer Nature
The registered company is Springer International Publishing AG
The registered company address is: Gewerbestrasse 11, 6330 Cham, Switzerland

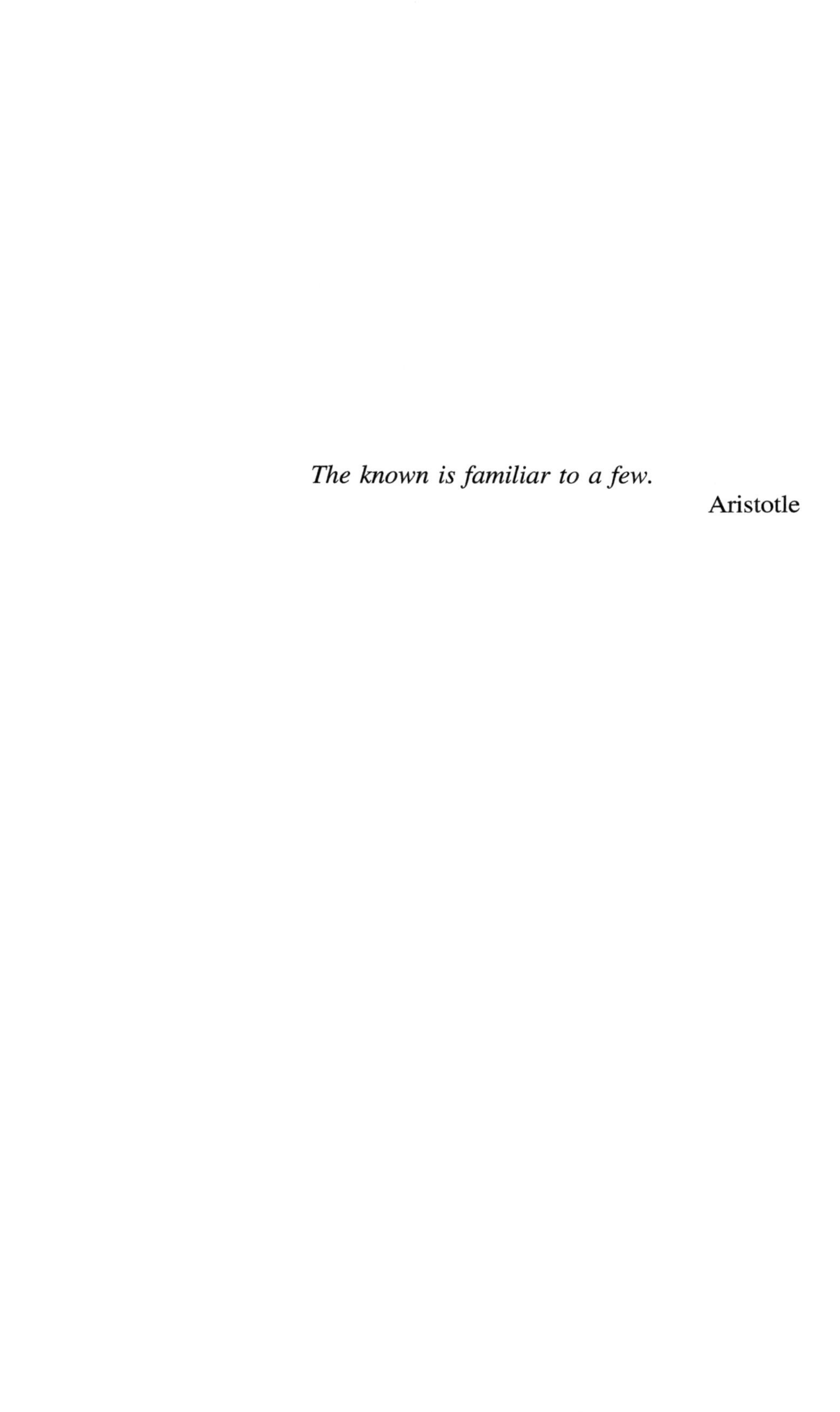

The known is familiar to a few.

Aristotle

Preface

The author of this textbook intends to consider all stages of the life cycle of the energy resources: extraction of mineral energy resources and mastering for power engineering renewable energy, transportation of mineral energy raw materials to the place of consumption, the conversion of primary energy sources into electrical and/or thermal energy, transportation and distribution among the customers, and energy storage (if necessary).

The author hopes that this approach—an analysis in the book of all main issues of the current state, problems and prospects for the most important sectors of the modern economy and daily life—power engineering—will allow the reader with minimal time and effort to get a holistic view of it. Naturally, such an approach does not allow all the problems to be considered deeply enough. This is the task of specialized textbooks and monographs. In these circumstances, all the questions in the textbook were conditionally divided into two groups. The former include the following issues: the impact of the energy sector on the socioeconomic and environmental conditions for the existence of the humanity, new and perspective primary energy sources (nonconventional renewable energy, auxiliary fuel resources), poorly developed new and promising ways of producing electricity and thermal energy (hydrogen energy, fast reactors, controlled thermonuclear fusion), accumulation of various types of energy (electrical, thermal, chemical, kinetic, and potential), small distributed power, and DC power transmission.

The latter include issues related to the well mastered in the industry and covered in the educational and scientific literature energy technologies: traditional production and transportation of energy resources (gas, oil, coal, and nuclear fuel); operation of powerful thermal, hydraulic, and nuclear power plants; transmission of electricity by means of three-phase current; and energy saving and energy efficiency at the enterprises of the fuel and power complex (FPC). Questions of the second group are considered in the book only concisely.

A great deal of attention given to main problems of power engineering includes the following: (1) exhaustion of nonrenewable power-generating resources ("Hunger for energy"), (2) negative impact of energy facilities on the environment ("Ecological infarct"), and (3) social and geopolitical threats.

The **first problem** is due to the rapid growth of the specific energy consumption (by each person) and the population growth on the planet. Today, it is particularly relevant for those countries that do not have their own energy resources in sufficient quantities and cannot buy it in abroad due to poverty [1, 2.]

Scientists studying the problem of sustainable development of human society believe that the **second problem**—the harmful impact of industry (first of all the objects of energy industry) on the environment—is of major concern today [3–5.]

Seriousness of **the third problem** "the social and geopolitical threats" is well expressed by journalists: "The current wars have not only the smell of the blood, but also the smell of oil" [6]. The reality of this problem confirmed quickly and easily— just compare the map of contemporary armed conflicts with the map of location of the main oil and gas reserves. This problem is not discussed in the textbook, as the true causes of today's military conflicts, their cause and results are covered in details by the mass media. Besides, the author of the textbook is not a professional politician and his personal point of view may not always reflect objective reality.

Methods for solving the first two problems based on mastering new production and energy consumption technologies, increasing the efficiency of utilization of power resources, and improving the life organization (life style) are analyzed. Upgrade strategy for power energy development and environment protection—one of the world trends in the past 1.5–2 decades. In its line is the Russian President decree "On the Strategy of the Russian Federation Scientific and Technological Development" signed in 2016. Important place is given to "New power engineering and environmental protection."

Many experts believe that the situation with growing demand for electric power for humanity can be improved only through revolutionary changes in the energy sector. However, on the path of reforms, there are a number of problems.

Factors determining the onset and rate of development of global energy revolution are

1. Economic—requirements in investment on a huge scale.
2. Technological—the inability of quick change of technological structures.
3. Social—(a) rapid population growth in developing countries and (b) the inability to quick change the existing domestic foundations.

A difficult task is provision of humanity with the electrical and thermal energy in developed countries. It solved by the fuel and power complex (FPC). It includes five main sectors (enumerated in accordance with technology in electrical power engineering):

- Search and investigation of fields (oil, gas, coal, shale oil and gas, and uranium);
- Fuel industry (production of oil, gas, coal, shale, peat);
- Pipeline transportation fuels (oil-trunk pipelines, oil products pipelines, and gas pipelines);
- Power production;
- Power transmission and distribution.

For example, in Russia, the FPC ensured about 35 % of the gross domestic product, more than 40% of the revenue item of the budget, and about 65–68 % of the currency income of Russia.

In the present century, one of the most important problems for humanity is the development of power engineering as a branch of science and economy without the damaging effects on the environment—this is the prevailing opinion among scientists around the world.

This statement is confirmed, including the fact that the International Global Energy Prize (~ 1 million $ by analogy with Nobel Prize) was established in Russia in 2003 on the initiative of Nobel laureate Zhores Alferov. It awarded for outstanding basic and applied research in the field of power engineering.

In the program ads of Administration of US President D. Trump in his inauguration day, the first task is energy.

Over the past century, the electric power industry continues to contribute to the welfare, progress, and technological advances of the human race. The growth of electric energy consumption in the world has been phenomenal. In the USA, for example, electric energy sale has grown well over 400 times in the period between the turn of the twentieth century and the early 1970s. This growth rate was 50 times as much as the growth rate of all other energy forms used during the same period.

The industry now faces new challenges associated with the interaction of power system entities in their efforts to make crucial technical decisions while striving to achieve the highest level of human welfare. This issue is discussed in Chap. 1 and in the introductory parts of several other chapters.

The second chapter provides the brief historical information on the formation and development of the electricity study and its application that makes it possible for students and young scientists to acquire the skills of perception of special disciplines in their historical correlations with fundamental science.

Each chapter concludes the questions and tasks to help reader focus attention on key issues and consolidate the new knowledge. The reference list includes only the English-language publications that are easy to find, and that contain materials sufficient for in-depth study of the problems discussed in the textbook. (The exception is made only for two monographs: [1] Bauman Z. Leben in der Fluechtigen Moderne.—Frankfurt am Main, 2007 (in Germ.); [35] Schwab A. J. Elektromagnetische Vertraglichkeit (Vierte, Neu Bearbeitete Auflag), Springer-Verlag Berlin Heidelberg, 1996 (In Germ.)

I wish to thank Ph.D. Shamanaeva L.G. for taking part in the translation of the book and leading engineer Bogdanova E.V. for her help in preparing the manuscript.

With gratitude will be accepted comments and suggestions that readers can send to the e-mail address: vyush@tpu.

Tomsk, Russia Vasily Y. Ushakov

References

1. Bauman Z. Leben in der Fluechtigen Moderne [in German]. – Frankfurt am Main. 2007.
2. Meadows DH, Meadows D, Anders J, Berens I. The limits to growth: the 30-year update, London-Sterling, VA: Earthscan; 2004.
3. Weizsaecker E, Hargroves K, Smith M, et al. Factor 5: transforming the global economy through 80% increase in resource productivity, London, UK: Earthscan; 2009.
4. Lomborg B. The asceticale environmentalists: Measuring real state of the world. Cambridge: Cambridge University Press; 2001.
5. Weizsaecker E, Lovins A, Lovins L. Factor four. Doubling of wealth—halving recourse use. London: Earth scan Publications Ltd.; 1997. P. 400.
6. Economides M, Oligney R. The color of oil (The history, the money and the politics of the world's biggest business). Katy, TX: Round Oak Publishing Company; 2003.

Contents

Abbreviations

AAPP	Air accumulating power plant
AC OTL	Alternative current overhead transmission line
ACLF	Artificial compounded liquid fuel
APE	Atomic power engineering
APG	Associated petroleum gas
BM	Biomass
BS	Bituminous sandstone
CHPP	Combined heat and power plant
CIS	Commonwealth of Independent States
CPP	Condensation power plant
CS	Capacitor storage
CTF	Controlled thermonuclear fusion
DC OTL	Direct current overhead transmission line
DG	distributed generation
DPP	Diesel power plant
EC	Energy consumption
ECG	Electrochemical generator
EES	Electric energy storage
EMC	Electromagnetic compatibility
EMS	Electromechanical storage
EPE	Electric power engineering
EPS	Electric power system
ES	Energy storage
FACTS	Flexible Alternative Current Transmission System
FE/FC	Fuel element/ fuel cell
FNPP	Floating nuclear power plant
FNR	Fast neutron reactor
FPC	Fuel and power complex
GDP	Gas distributing point
GDS	Gas distributing station

GH	Gas hydrate
GPI	Gas piston installation
GTI	Gas turbine installation
GTPP	Geothermal power plant
HCS	Housing and commune sector
HDI	Human development Index
HPP	Hydroelectric power plant
HTS	High-temperature superconductivity
HWS	Hot water supply
IAEA	International Atomic Energy Agency
ICE	Internal combustion engines
IEA	International Energy Agency
IES	Inertial (dynamic) energy storage
IPS	Independent power supply
ITER	International Thermonuclear Experimental Reactor
ITF	Inertial thermonuclear fusion
LP NPP	Low-power nuclear power plant
MG	Micro-grids
MM	Mine methane
MOX fuel	Mixed oxide fuel (UO2 + PuO2)
NFC	Nuclear fuel cycle
NPP	Nuclear power plant
NRPG	Nontraditional renewable power generation
OPP	Osmosis power plant
OTEC	Ocean thermal energy converter
OTL	Overhead transmission line
PFBR	Prototype Fast Breeder Reactor
PGR	Power-generating resources
PSPP	Pumped storage power plant
RES	Renewable energy source
RP	Reactive power
RPS	Renewable power sources
SB	Storage battery (electrochemical storage)
SC	Slate coal
SGI	Steam and gas installation
SIES	Superconducting inductive energy storage
SNF	Spent nuclear fuel
SPP	Solar power plant
SSP	Silicon solar panel
TdPP	Tidal power plant
TEG	Turbo-expander generator
TNR	Thermal neutron reactor
TPP	Thermal power plant
UDTL	Underground distribution transmission line
UEPS	United Electrical Power System

UHV TL	Ultra high-voltage transmission line
UNFCCC	United Nations Framework Convention on Climate Change
URPS	Unconventional renewable power sources
WavePP	Wave power plant
WEC	World Energy Conference
WTU	Wind-turbine unit
WWER	Water–Water Energetic Reactor

Chapter 1
Power Engineering as a Basis for Progress of Civilization

The first chapter in this textbook should provide insights into the influence of power engineering on the present day civilization. It is through an understanding of the past developments and achievements we can understand our present and forge ahead to future advances.

1.1 Main Concepts and Definitions

Before we briefly consider the history and current problems of electric power engineering (EPE) as part of the FPC and challenges to all human community associated with power engineering, we recall the main concepts and definitions relating to this issue.

Energy (action or activity in the Greek language) is a common quantitative measure of various forms of motion of matter.

The main properties of energy are:

- Ability to manifest itself only when states or positions of various objects in the surrounding world are changing,
- Each form of energy can be transformed into another form,
- Possibility to be quantified,
- Ability to display effects useful for a human.

Depending on the nature, there are eight basic forms of energy: mechanical, thermal, electrical, chemical, magnetic, electromagnetic, nuclear, and gravitational.

There are numerous ways to classify the types of *power-generating natural resources*. They include the source of origin, the state of development, and the renewability of the resources. In articles and monographs, you can find different names to the same types of energy resources, as well as their different classifications. Here are the most common. For example, natural resources can be combined according to of their origin into the following types:

© Springer International Publishing AG 2018
V.Y. Ushakov, *Electrical Power Engineering*, Green Energy and Technology,
https://doi.org/10.1007/978-3-319-62301-6_1

Biotic or "biomass" (BM)—these terms include a wide range of materials of biological origin, namely:

(1) Natural vegetation products (wood, wood waste, peat, leaves, etc.);
(2) Wastes, including flammable, household, and industrial waste;
(3) Agricultural residues (stalks, leaves, manure, etc.);
(4) Specially grown high-yielding agricultures.

The chemical composition and formation processes of the traditional "fuel triad" (coal, oil, and gas) is similar to that of biomass. However, fossil fuels formed millions of years ago, and therefore they are not considered as renewable energy sources.

Abiotic: these resources come from nonliving and organic materials. Examples of these resources include fuel for nuclear reactors—uranium, thorium (in a future) and thermonuclear reactors—helium, deuterium, and tritium.

Natural resources can categorize based on their stage of development including:

- ***Potential resources***: these resources that exist in a region and may be used in future. For example, if a country has petroleum in bowels of the earth, it is a potential resource until they will be extracted.
- ***Actual resources***: these are resources that have been surveyed, their quantity and quality has been determined, and they are currently being used. The development of actual resources depends on technology.
- ***Reserve resources***: this is the part of an actual resource that can be used profitably in future.
- ***Stock resources***: these resources are explored, but can not be used at present due to the lack of appropriate technologies. An example of a stock resource is hydrogen. There is constant worldwide debate regarding the allocation of natural resources. The discussions are centered around the issues of resource depletion and the exportation of natural resources as a basis for many economies (especially developed nations). The vast majority of natural resources are exhaustible which means they are available in a limited quantity and can be fully depleted if their use is not managed correctly. Natural resource economics aim to study resources in order to prevent depletion.

Energy extracted directly from the nature called ***primary energy***.

Energy produced by a human by primary energy conversion in power plants is called ***secondary energy*** (electrical, vapor, and hot-water energies). Thermal, chemical, and nuclear energies in the literature sometimes are called ***internal energy***, that is, energy stored inside a substance. The modern science admits the existence of other types of energy that have not yet been discovered. ***Power-generating resources (PGR)*** are material objects containing energy suitable for its utilization by a human at present or in the nearest future. Depending on sources, PGR subdivided into primary (natural) and secondary (subsidiary) ones.

In turn, ***primary PGR*** subdivided into ***fuel*** and ***non-fuel*** according to methods of their utilization.

1. Natural resources also classified based on their renewability:

   - ***Renewable natural resources***: these resources can replenish. Examples of renewable resources include sunlight, wind, water flow, and others. They are available for a long time and their quantity does not depend on consumption. However, renewable resources do not have a rapid recovery rate, but some of them can be depleted if overused.
   - ***Non-renewable natural resources***: these resources are formed extremely slowly in the scale of human life and even the life of generations. It is reflected in their name. Resources are non-renewable if their consumption rate exceeds the recovery rate. For example, fossil fuels are non-renewable natural resources.

2. According to their dynamics, they are resources *participating in the permanent energy cycle* (solar and space energy and so on), *deposited* resources (coal, oil, gas, and so on), and *artificially activated* resources (nuclear and thermonuclear energy).

3. According to the effect on the Earth and near-Earth space, they *add* or *do not add* energy to the natural planetary energy influx.

4. According to the economic efficiency, they are: *gross (theoretical)* resources encompassing the whole energy of the given energy resource, *technical* resources incorporating the energy that can be extracted from the given resource at the existing technical level, and *economic* resources encompassing the energy whose production from the given resource type is economically profitable for the existing costs of equipment, materials, and labor. The relative contribution of the last two resources increases in the process of refinement of power equipment and mastering new technologies.

Secondary PGR (SPGR) are power by-products unused in the given production process or facility. The secondary PGR characterize the energy potential of products, wastes, by-products, and intermediate products of technological aggregates (systems) that cannot will be used in this aggregate but can be used partially or completely to supply power to other consumer.

Chemically bound heat of products of fuel refining (oil refining, gas-generator, coking, coal-enrichment, and so on) systems does not belong to the SPGR. Thermal energy of wastes used for heating flows entering the aggregate—the secondary PGR source (for regeneration or recuperation)—belongs to them.

The amount of heat, cold, or electrical energy produced from the SPGR in a utilization system are called *a yield*. The yield can be possible, economically expedient, planned, and actual. The difference is understandable from the definitions.

By nature, the SPGR are subdivided into *thermal* and *combustible* and an *excessive pressure*.

The *thermal* SPGR incorporate physical heat of outgoing gases, main products and by-products, ash and slag, and hot water and vapor used in technological systems and working bodies of cooling systems of these technological systems.

When the thermal SPGR in the form of heat of outgoing gases of technological aggregates and heat of main products and by-products are properly utilized, much fuel is saved. Calculations demonstrated that the cost of heat energy produced in utilization systems is lower than expenses on generation of the same amount of heat energy in main power plants.

The *combustible* SPGR incorporate combustible gases and by-products that can be used as fuel in other systems and cannot be used in the given technological process, for example, by-products of timber industry (chips, sawdust, shavings, and cuttings), combustible parts of buildings and structures destroyed because of unfitness for further exploitation. Others solid or liquid fuel by-products are used in pulp and paper production.

The SPGR *of excessive pressure* incorporate potential energy of gases, water, and vapor outgoing from a system and having an excessive pressure that still can be utilized before emission into the atmosphere or discharge into water reservoirs, tanks, or other containers.

The excessive kinetic energy also refers to secondary power resources of excessive pressure.

The electrical power system (EPS) is a subsystem of the fuel and power complex (FPC) that, in its turn, consists of following three systems:

- the source of energy—generation (production)
- bulk transfer—transmission
- supply to individual customers—distribution.

It is a set of power stations, step-up and step-down transformer substations, and transmission and distribution lines.

Figure 1.1 illustrates this set of systems together with their direct and reverse links (shown by dashed straight lines). This figure and the above definition emphasize the *system approach* to power engineering in the context of which it is considered as a large system incorporating parts of large systems as subsystems.

The development and functioning of power engineering depends on social and demographic factors (political and economic states of the given country, labor resources in it, population accommodation, location of power sources, and so on). The volume of energy production is connected with its consumption in industry, agriculture, everyday life, and transport.

Since power of artificial power systems and natural geophysical processes that influence our planet become comparable, power engineering will play more and more important role in the future of humanity. This is not only demographic and ecological but the social and political factors that influence international affairs. Essential factor influencing international cooperation, conflicts, and agreements between governments is the extreme uneven distribution of energy resources on the planet, Fig. 1.2.

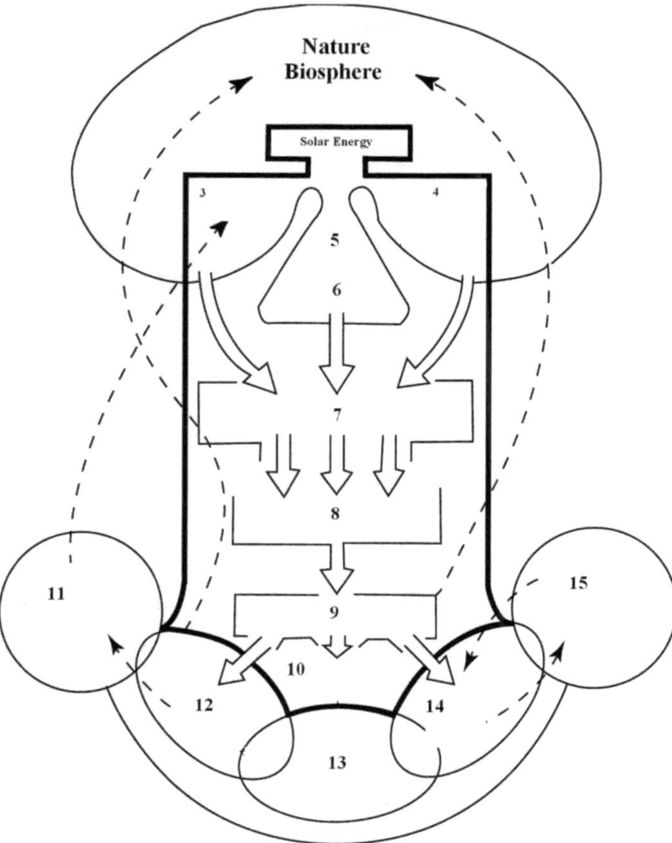

Fig. 1.1 Power engineering and related subsystems: *1* nature. Biosphere; *2* solar energy; *3* organic fuel; *4* hydraulic energy, geothermal energy, tidal energy; *5* photosynthesis; *6* nuclear energy; *7* energy production; *8* energy conversion; *9* energy transmission; *10* conversion and distribution of energy; *11* development of country. Demographic factors; *12* consumption in industry; *13* consumption in a transport; *14* consumption in life; *15* development of culture. Social factors

It is very heterogeneous across countries and per capita energy consumption. The difference in the electricity consumption per capita in different countries reaches 50-fold.

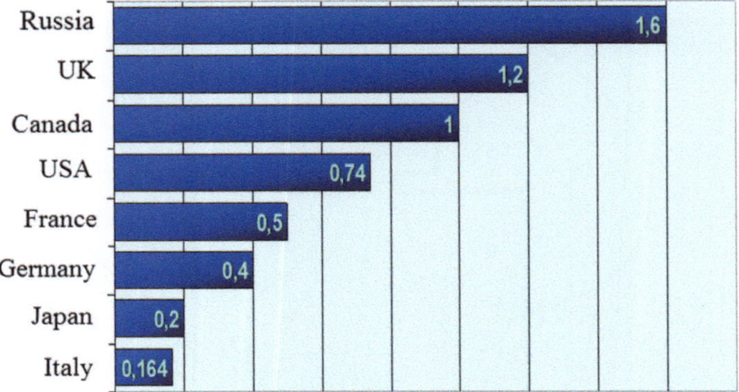

Fig. 1.2 Provision in energy resources of G-8

1.2 Influence of Power Engineering on the Development of Humanity

Historically, the process of energy consumption on our planet was strong non-uniform. Figure 1.3 illustrates this process. Humanity has spent approximately 900–950 thousand TWh of energy of all types during its lifetime. Moreover, about 2/3 of this amount has been spent in the last 30–5 years.

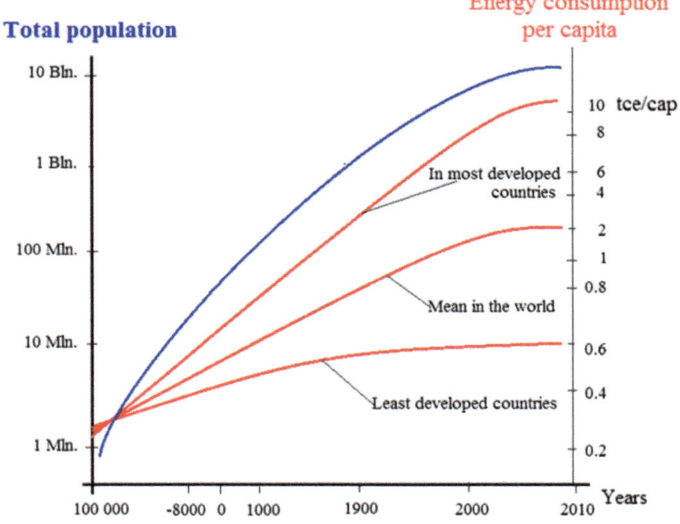

Fig. 1.3 Growth of energy consumption (per capita) and population in the world

Fig. 1.4 Dynamics of world energy consumption from 1990 up to 2035, 10^{15} BTU

Figure 1.3 shows that especially rapidly the energy consumption has been increasing since the 20th century. Dynamics of world energy consumption from 1990 up to 2035 is shown in Fig. 1.4.

Annual growth in world energy consumption up to 2020 will amount 1.6–2% and then slows down to 1%; in 2035, the world will consume about 17.4 billion tons of oil equivalent. This is by 33% higher than in 2011.

The non-uniformity of energy consumption is also typical over this period. Thus, each human in the prehistoric epoch used his/her muscular force and the energy of fire and spent nearly the same amount of energy. Hence, the energy distribution can be considered approximately uniform (1:1). At present, the non-uniformity of energy consumption per capita is huge. In different countries, it is expressed by the 1:40 ratio. The non-uniformity of electrical energy consumption is even higher. Thus, about 25,200 kWh of electrical energy is spent per capita in Norway, whereas in Ethiopia only 54 kWh is spent per capita.

Even in a small group of countries that produce and consume the greatest amount of electricity there is a significant difference in the value of this index, Table 1.1.

Progress in power engineering and in electrical consumption systems is of great importance, because they increase the labor efficiency and hence the gross output [1]

Figure 1.5 shows the dependence of the gross national population income (NI) on the energy consumption (EC). The income expressed in dollars USA and energy in Btu was recalculated per capita (Btu is the British unit of heat equal to 1.055 kJ).

Table 1.1 Electricity generation and consumption of leading countries in 2011

Country	Electricity TWh per year	
	Consumption	Production
China	4693	4604
India	600.6	835.3
USA	3741	3953
Russia	857.6	925.9
Japan	859.7	930

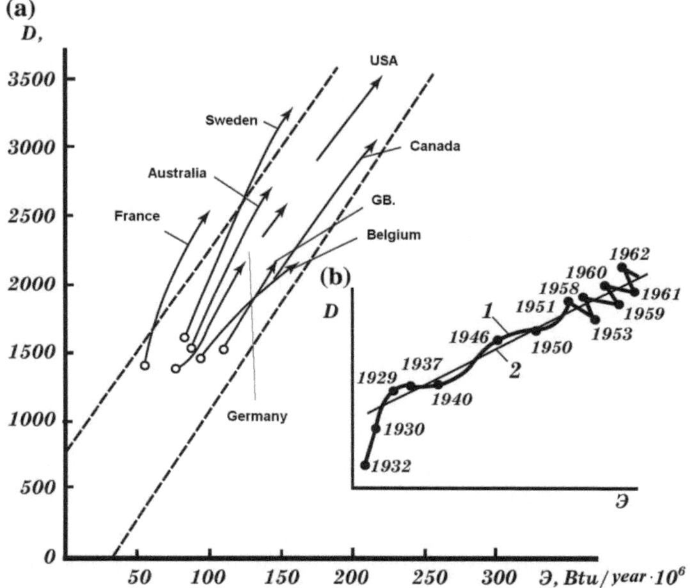

Fig. 1.5 Correlation between the energy consumption and the gross income

In Fig. 1.6 shows the connection of other indicators of well-being of residents of a country—Human Development Index (HDI)—with the volume of electricity consumption.

The period of scientific and technical revolution and formation of postindustrial society in the most developed countries differs qualitatively from the preceding stages of development. This difference is primarily a huge revolutionary advance of productive forces and creation of technically perfect and highly efficient automated means of production on a large scale.

A scientific and technical revolution is a complicated process with diversified social consequences. Progress in science and technology is determined by the unity of evolutionary and revolutionary changes. Moreover, with the corresponding

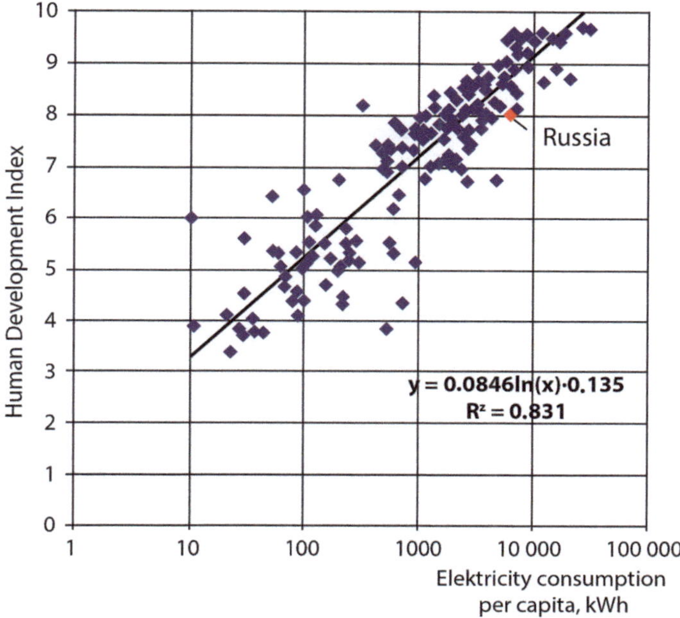

Fig. 1.6 There is connection the HDI with the volume of electricity consumption for different countries (For Russia marked by a red sign HDI = 8–60th place out of 173 countries.)

premises, the effect of internal mechanisms in the development of science and technology and of public needs can transform the evolutionary changes into revolutionary ones. Any technical revolution is characterized by radical changes in production means or production process itself. It involves even branches of industry, and as a result, a qualitatively new material and technical base are created.

Considering the current scientific and technical revolution and processes in the postindustrial society, we must take into account the history of technical development, basic technical achievements, and scientific discoveries over the past few years. This allows proper directions of scientific and technical progress chose that will not only ensure fast progress, but also will completely meet the requirements of the present and future generations.

Advances in automation and electrification of the production process and means of transport are also important for present and future revolutionary changes in power engineering and the gross income. Human consciousness and mind evolve in the process of modernization of production means and forces as people learn to change the nature. Attempting to penetrate the mysteries of nature, humans are making efforts to use capabilities of nature to meet its own needs. Such natural phenomena as a lightning, solar energy, rise and fall of tides, and many others that were perceived as mysterious forces by ancient people became more understandable in the course of time. The worship of these and elemental forces makes people to idolize them. It was no mere chance that the Sun was a supreme god for many ancient people. This naive idea of the Sun evaluates correctly its significance as a source of almost all energy used by humans and as a source of life. Solar heat was the first energy source used by a human. A legend about Prometheus reported the greatest event in the human life, namely, that people had been learned to make and to support fire and hence to utilize chemical energy accumulated in an inorganic fuel. This form of energy is widely used at present.

Mastering of natural PGR has stimulated the creation of machines capable of execution of complex operations, and a considerable portion of physical and a certain portion of intellectual work was thrown on these machines. Their modernization gave people free time for the most creative work and better understanding of laws of nature to use them for the benefit of humanity. In turn, this promoted the creation of even better instruments of production.

The demands for energy permanently increased sending people in a search for new energy resources and new methods of energy transformation from one type into another. At present the use of solar energy, chemical energy of organic fuel, mechanical energy of water in rivers, seas, and oceans, wind energy, and nuclear energy liberated in the process of fission of heavy nuclei becomes already conventional. The use of thermonuclear energy of synthesis of light elements is very promising; this will solve the problem of energy supply arising in connection with exhaustion of organic fuel reserves in near future.

Returning to the question of the development of power engineering, it should be noted that explosive progress in technique and its modern level would be impossible without qualitatively new energy types, first, electrical energy. *Electrical energy provides the basis for modern civilization.* Without electrical energy, the normal life of modern society is impossible. Exactly electrical energy was a driving force that led to large-scale machine manufacture and ensured unprecedented progress in the development of productive forces. Electric energy can be easily transmitted over a long distance and transformed into other energy types.

Recently, the relationship of power engineering with *the biosphere*, that is, the space in which all living beings exist, becomes more clearly pronounced. This happens on our planet because the power and energy that a human has learned to produce artificially become comparable to natural ones (See Chaps. 5 and 8).

Questions and Tasks to Preface and Chapter 1

1. What are types of energy most widely used by humanity?
2. Briefly explain notions of the primary and secondary energy and name their main types.
3. What is the connection between the level of social development (Human Development Index) and the per capita energy consumption?
4. How does energy supply actually affect the human daily live?
5. What are the consequences of the extremely uneven distribution of energy resources across countries and regions?

6. Why do you think the overpopulation of the Earth can be very dangerous for the future of humanity?
7. Why some experts do not treat seriously the threat of "hanger for energy"?
8. How do you understand "energy saving"?
9. What are the three problems connected with the power production and consumption?
10. Why did the oil price on the international market decrease during 2014–2015?
11. What is the long-range prediction of the hydrocarbon price of most analysts?
12. How can be seen the influence of climate in concrete country on the power security?

Reference

1. Glachant J-M, Finon D, de Hauteclocque A. Competition, contracts and electricity markets. A new perspective. EE Publishing; 2011.

Chapter 2
From the History of Electrical Power Engineering

This chapter describes a brief perspective of development of electric power systems. This is not a detailed historical review, but rather it uses historical landmarks as a background to highlight the features and structure of the modern power systems discussed in the next chapters [1]. (Fragments of the newest history can be found in [2].)

2.1 Formation of Electrical Engineering as an Independent Engineering Branch (1870–1890)

Electrical equipment systems were used only in laboratories, and consumers had no sufficiently high-power and economically efficient source of electric energy. Such a source was created in 1870. The next 15–20 years passed as years of creation of main electrical engineering devices for industrial and household applications and formation of new engineering branch—electrical engineering.

The experience in application of electromagnetic telegraph suggested the principal feasibility of transmission of significant amount of electric energy using conductors. Already in the 1840–1850s electrical engineers from the USA, Italy, and some other countries came up with the idea of an electrified railroad equipped with power transmission at long distances. Experiments of I. Fontaine (1833–1910), a French electrical engineer, became world-known.

The international exhibition from which the history of power transmission began was held in Vienna in 1873. In this exhibition, I. Fontaine demonstrated reversibility of electrical machines. A cable more than 1 km long connected a generator and an engine. The engine actuated a pump of an artificial decorative waterfall. This experiment demonstrated the principal feasibility of *power transmission* at long distances. At the same time, I. Fontaine was not convinced of an economic efficiency of power transmission, because he observed significant decay

© Springer International Publishing AG 2018
V.Y. Ushakov, *Electrical Power Engineering*, Green Energy and Technology,
https://doi.org/10.1007/978-3-319-62301-6_2

of the engine power when he connected the cable and large energy losses in the cable. W. Stanley in the US at Great Barrington, Massachusetts, installed the first practical AC distribution system in 1866, which acquired the American rights to the transformer from its British inventors Goliard and Gibbs. Early AC distribution utilized 1000-V overhead lines. The Nikola Tesla invention of the induction motor in 1888 helped to replace DC motors and fastened the advance in use of AC systems. The first American single-phase AC system was installed in Oregon in 1889. Southern California Edison Company established the first three-phase 2.3 kV system in 1893. By 1895, Philadelphia had about twenty electric companies with distribution systems operating at 100-V and 500-V two-wire DC and 220-V three-wire DC, single-phase, two-phase, and three-phase AC with frequencies of 60, 66, 125, and 133 cycles per second, and feeders at 1000–1200 V and 2000–2400 V.

As is well known, the losses in a transmission line depend on voltage and wire resistance, and cross sectional area. It is virtually impossible to decrease the resistance of wires, since copper that became the main material for manufacturing wires has extremely low resistance. Only nowadays researchers are involved in theoretical and experimental works aimed at reducing the resistance of transmission lines based on the superconductivity phenomenon (cryogenic transmission lines). *Therefore, there were only two methods of decreasing losses in the transmission line, namely, to increase the wire cross-section or to increase the voltage.*

The first method was intensively studied in the 1870s, because an increase in the cross sectional area of conductors seemed more natural and easier for realization than an increase in the voltage. However, numerous attempts of implementation of this method have finally demonstrated the lack of its prospects.

The deadlock character of the method for decreasing losses by increasing the cross sectional area of wires was rather quickly realized.

The third method of solving the problem of electric energy transmission had been comprehended theoretically for a long time.

M. Deprez (1843–1918), a French engineer (later Academician) and D.A. Lachinov (1842–1902), Professor of Physics of the St. Petersburg Forest Institute, carried out the most comprehensive studies of this problem independently in 1880.

The report of M. Deprez entitled "On the Efficiency of Electric Engines and Measurement of the Amount of Energy in an Electric Circuit" was published in minutes of the Paris Academy of Sciences in March 1880. *M. Deprez is known in the history of electrical engineering as an inventor of several systems of an ammeter, wattmeter, deadbeat galvanometer, and principle of mixed (compound) excitation of electric machines, electromagnetic hammer (reciprocating engine), and electrical motion synchronization system.*

In his report, M. Deprez mathematically proved that the efficiency of a system comprising an electric motor and a transmission line was independent of the line resistance. This conclusion seemed paradoxical for M. Deprez, since from the very beginning he failed to establish that the increase of the line resistance had no effect on the efficiency of electric power transmission only under a specific condition, namely, when the transmission voltage increased.

These conditions were first pointed out by D.A. Lachinov in his paper entitled "Electromechanical work" published in June 1880 in the first issue of the Journal *Electricity*. Based on mathematical operations, he demonstrated that the efficiency of electric energy transmission is independent of the distance only under condition of increasing rotation rate of the generator (that is, when the voltage in the line increased, because the EMF produced by the generator is proportional to its rotational speed.) D.A. Lachinov also determined quantitative interrelationships among the parameters of the transmission line and proved that to keep the transmission efficiency constant when the resistance increases n folds, the rotational speed of the generator should be increased by square root of n times.

In 1882 M. Deprez built the first Misbach–München transmission line 57 km long. A steam engine actuating a 3-hp dc generator producing a dc voltage of 1.5– 2 kV was placed at one end of the test transmission line in Misbach. The energy was delivered through steel telegraphic wires 4.5 mm in diameter to the territory of the München exhibition where the same machine operated as an electric motor that actuated a pump of an artificial waterfall. Though this first experiment gave no favorable technical results (the transmission efficiency did not exceed 25%), its significance could not be underestimated, because the electric energy transmission from Miesbach to München was the starting point for the development of methods and means of electric energy transmission at long distances.

In 1885 new experiments on power transmission between Crale and Paris at a distance of 56 km were conducted. Specially constructed machines that yield the output 6 kV were used as dc high-voltage generators. The mass of this machine was 70 t, its output power was about 50 hp, and its transmission efficiency was about 45%.

During the same period, case transmissions of electric power at long distances were performed for industrial applications with the efficiency as great as 75%.

Nevertheless, attempts to transmit the electric power using the direct current carried out in the 80s did not bring desirable results. In this case, it is important to emphasize the arising contradiction. On the one hand, the design and production of dc electric machines and systems had gained wide acceptance and the dc engines had good performance characteristics that met the majority of industrial requirements. Therefore, there were no severe obstacles for electrification of the industrial stock of machines. However, on the other hand, large-scale industrial electrification could be realized only in the case of centralized production of electric power and hence only when it could be transmitted at long distances.

Difficulties connected with dc energy transmission set scientists thinking over the theory and technology of alternating current.

Once the main parts of ac systems (generators and transformers) had been developed, attempts of industrial ac power transmission were undertaken. In 1883 L. Golyar transmitted a power of 20 hp at a distance of 23 km to supply the illumination system of the London Subway. Transformers stepped up the voltage to

1500 V. Next year L. Golyar demonstrated transmission of a power of ~ 40 hp. at 2 kV at a distance of 40 km in the Turin Exhibition.

However, in the second half of the 1880s a problem of connecting a electric motive load into an electric circuit arose that troubled engineers and scientists. Thus, the transmission of electric power in single-phase ac circuits involved contradictions no less serious than the dc transmission. The voltage in single-phase ac circuits could easily be stepped up and down in any desirable limits using transformers. Therefore, there were no difficulties in the electric power transmission. However, single-phase ac engines had characteristics unsuitable for practical applications. In particular, they either had no starter at all (synchronous engines) or were hard to start up because of difficult conditions for current commutation (collector engines). Therefore, the application field of single-phase ac circuits was limited almost exclusively by illumination systems, which certainly could not meet the requirements of the industry.

Whereas N. Tesla and his employees attempted to improve the two-phase system, a better three-phase electric system had been developed in Europe. In 1887–1889, scientists and engineers developed multiphase systems more or less successfully.

M.O. Dolivo-Dobrovolskii (1862–1919), who managed to assign the practical character to his work, was most successful in the development of multiphase systems. Therefore, he is correctly considered to be a founder of three-phase system technology.

M.O. Dolivo-Dobrovolskii was born in St. Petersburg, studied in Riga Polytechnic Institute, and completed his education in Higher Technical School in Darmstadt (Germany). In this School much attention was given to practical applications of electricity.

M.O. Dolivo-Dobrovolskii developed a three-phase electric system and a perfect design of the asynchronous motor that still remained unchanged.

The first important step made by him was the invention of a rotor with a winding reminiscent of a squirrel wheel the design of which remained virtually unchanged up to now.

The next step of M.O. Dolivo-Dobrovolskii work was the replacement of the two-phase system by a three-phase system. The first three-phase asynchronous motor with an output power of about 100 W was constructed in spring of 1889. This motor was supplied by the current from a three-phase single-armature converter and passed tests quite satisfactorily.

After the first single-armature converter he created the second high-power modification. Then manufacturing of three-phase synchronous generators began. Designs of engines constructed by M.O. Dolivo-Dobrovolskii were so perfect that remained virtually unchanged for more than 100 years of their existence.

The three-phase system became so widespread from the first years of its construction because it did solve successfully the problem of power transmission at long distances. However, the power transmission is efficient at high voltage, which for the ac can be obtained using a transformer. The three-phase system presented no principal difficulties for energy transformation, but required three single-phase transformers instead of one. Such an increase of the number of rather expensive transformers sent M.O. Dolivo-Dobrovolskii in search for a better solution. M.O. Dolivo-Dobrovolskii invented a three-phase transformer. In October 1891 he submitted an application for a patent on a three-phase transformer with parallel rods arranged in one plane. This design is still retained.

Works devoted to three-phase circuits were also aimed at improving the power transmission characteristics. To adjust voltages in separate phases and to have two voltages in the system, namely, phase and linear voltages, M.O. Dolivo-Dobrovolskii suggested a four-wire scheme for three-phase circuit or in other words, *a system with a neutral wire* in 1890. Simultaneously he indicated that the grounding could be used instead of neutral or zero wire. Thus, all basic elements of the three-phase system of power supply including a transformer, three- or four-wire transmission line, and an asynchronous motor in two main modifications (with phase and cage rotors) was constructed for 2–3 years. From all possible designs of multiphase synchronous generators the principle of construction of which had already been known, only three-phase machines have received wide application. In such a way, the three-phase electric current system came into being and received wide recognition.

M.O. Dolivo-Dobrovolskii recommended a sine curve as a main current waveform. He also recommended a current frequency of 30–40 Hz. However, only two mains frequencies were chosen after a critical analysis: 60 Hz in the USA and 50 Hz in other countries. These frequencies were optimal, because an increase in the mains frequency led to an excessive increase of the rotation speed of electric machines (with the same number of poles), whereas a decrease in the mains frequency adversely affected the uniformity of illumination.

2.2 Next Stages of Power Engineering—Sustainable Formation and Development

Electrification in the history of science and technology dates back to 1891 when a three-phase electric transmission system was demonstrated and tested in the International Electrical Engineering Exhibition in Frankfurt am Main.

In August 1891, 1000 incandescent lamps supplied by a current from the Laufen hydroelectric power station were first lit in the exhibition; on September 12, 1891 the M.O. Dolivo-Dobrovolskii engine put into operation a decorative waterfall.

What was this first three-phase system of electric transmission like?

In the Laufen hydroelectric power station, the electric power generated by a turbine was transmitted through a conical gear to the shaft of a three-phase synchronous generator (with an output power of 230 kV A, rotation speed of 150 rpm, and voltage of 95 V in which windings were connected in star). In Laufen and Frankfurt there were three three-phase transformers with a prismatic magnetic core. The transformers were immersed in tanks filled with oil.

The three-wire transmission line supports were made from wood with an average span of about 60 m. Pin-type porcelain-oil insulators carried copper wires 4 mm in diameter. An interesting feature of this transmission line was the installation of fuses on the high-voltage side: a segment 2.5 m long consisting of two copper wires each having a diameter of 0.15 mm was connected in the gap of each wire at the beginning of the line. To switch off the line in Frankfurt, a simple device was used to short the three-phase circuit. The fuses blew, the turbine started to increase its speed, and when the machine operator noticed this, he stopped the machine.

In the exhibition site in Frankfurt, a step-down transformer was placed from which 1000 incandescent lamps arranged on a huge board were supplied at a voltage of 65 V. The Dolivo-Dobrovolskii three-phase asynchronous motor that drove a hydraulic pump with an output power of about 100 hp that put into operation a small artificial waterfall was also demonstrated in this exhibition. Simultaneously with this high-power motor, M.O. Dolivo-Dobrovolskii exhibited a 100-W three-phase asynchronous motor with a fan fitted to its shaft and a 1.5-kW motor with a dc generator fitted to its shaft.

Tests of transmission system conducted by the International Commission gave the following results: the minimum electricity transmission efficiency (the ratio of the power on the secondary terminals of the transformer in Frankfurt to the power on the turbine shaft in Laufen) was 68.5%, the maximum efficiency was 75.2%, the linear voltage during the tests was about 15 kV and increased to 25.1 kV at a higher voltage, and the maximum efficiency was 78.9%.

These tests not only demonstrated the capabilities of electrical energy transmission but also resolved the old debates: the alternating current conquered the direct current.

The creation of the three-phase system was the most important stage in the development of electric power engineering and electrification. When the Frankfurt exhibition was over, the electric power station in Laufen became the property of Halbern situated at a distance of 12 km from Laufen. It was put into operation in early 1892. The electric power was used to supply all urban lighting mains as well as a number of small plants and workshops. The step-down transformers were established directly for consumers.

In 1892, the Bulakh–Erlikon line was put into operation in Switzerland. A hydroelectric power station with three three-phase generators having a power of 150 kW each was built at a waterfall in Bulakh. The electric power was transmitted at a distance of 23 km to supply a plant. After these first stations, a number of electric power stations were built rather fast. Most of them were built in Germany.

In America, the first three-phase station was built in the late 1893 in California. The hydroelectric power station had two generators with power of 250 kW each and two transmission lines at a generated voltage of 2500 V. The first line (12 km long) delivered energy for lighting, and the second line (7.5 km long) supplied a three-phase asynchronous motor with a power of 150 kW.

At the beginning, the rate of use of three-phase systems in America was much less than in Europe. This is explained by the fact that one of the largest American firms—the Westinghouse Electric Corporation—persistently attempted to organize building of the Tesla two-phase electric power stations and electric networks.

The American firm General Electric, the main competitor of the Westinghouse Electric Corporation, fast redirected and in counterpoise to the competitive firm organized building of three-phase systems. The Westinghouse Electric Corporation lost the competition.

Attempts of combining outdated and new engineering solutions were characteristic of transition periods of any engineering branch. Thus, during almost two decades since 1891 attempts were undertaken to combine three-phase systems with other systems. In these years there were electric power stations having generators of constant and alternating single-phase current together with two-phase and three-phase generators or any their combination. The voltages and frequencies were also different, and the consumers were supplied through separate lines. Attempts to use outdated systems together with electric equipment mastered by plants resulted in the creation of hybrid systems.

Two three-phase electric power stations were first combined in 1892 in Switzerland. In Russia, the first enterprise with three-phase power supply was the Novorossiisk elevator. This was a huge set of buildings, and the problem of power distribution over floors and buildings could be solved in the best way only with the help of electricity.

Drawings of three-phase machines were ordered to the Swiss Brown-Bowery Plant in summer of 1892. In 1893 the elevator was electrified. It is interesting to note that all machines designed abroad were manufactured in workshops of the elevator. Four synchronous generators having power of 300 kV A each were mounted in the electric power station built near the elevator. Thus, the total power of the electric power station was 1200 kW A, that is, it was the most powerful three-phase electric power station in the world at that time.

Three-phase engines having a power of 3.5–15 kW worked in elevator rooms. They actuated various machines and mechanisms. A portion of energy was used for illumination.

In conclusion it is necessary to say that the historical evolution of electrical engineering can be attributed to the work and discoveries of the scientists, mathematicians, and physicists who were not referenced throughout the text, but included in the following list and Table 2.1.

Charles A. Coulomb (1736–1806), French engineer and physicist, published the laws of electrostatics in seven memoirs to the French Academy of Science between 1785 and 1791. His name is associated with the unit of charge.

James Watt (1736–1819), English inventor, developed the steam engine. His name is used to represent the unit of power.

Alessandro Volta (1745–1827), Italian physicist, discovered the electric pile. The unit of electric potential and the alternate name of this quantity (voltage) are named after him.

Hans Christian Oersted (1777–1851), Danish physicist, discovered the connection between electricity and magnetism in 1820. The unit of magnetic field strength is named after him.

Table 2.1 Short prehistory of electrical engineering formation

Most important events	Years	Authors
Invention of an inductor—the simplest transformer with a disclosed magnetic core	1848	G. Rumkopf (1870–1890)
Demonstration of the long-range power transmission possibility	1873	I. Fontaine (1833–1910)
Development of the method of decreasing power losses in transmission lines by using high voltage	1880	M. Deprez (1843–1918) D.A. Lachinov (1842–1902)
Creation of the first 57-km transmission line (Misbach–Munich)	The last quarter of the 19th century	M. Deprez
Creation of three-phase systems	1885–1892	M.O. Dolivo-Dobrovolskii (1862–1919)

Georg Simon Ohm (1789–1854), German mathematician, investigated the relationship between voltage and current and quantified the phenomenon of resistance. His first results were published in 1827. His name is used to represent the unit of resistance.

Michael Faraday (1791–1867), English experimenter, demonstrated electromagnetic induction in 1831. His electric transformer and electromagnetic generator marked the beginning of the age of electric power. His name is associated with the unit of capacitance.

Joseph Henry (1797–1878), U.S. physicist, discovered self-induction around 1831, and his name has been designated to represent the unit of inductance. He had also recognized the essential structure of the telegraph, which was later perfected by Samuel F. B. Morse.

Carl Friedrich Gauss (1777–1855), German mathematician, and **Wilhelm Eduard Weber** (1804–1891), German physicist, published a treatise in 1833 describing the measurement of the Earth's magnetic field. The gauss is a unit of magnetic field strength, while the weber is a unit of magnetic flux.

James Clerk Maxwell (1831–1879), Scottish physicist, discovered the electromagnetic theory of light and the laws of electrodynamics. The modern theory of electromagnetics is entirely founded upon Maxwell's equations.

Ernst Werner Siemens (1816–1892) and **Wilhelm Siemens** (1823–1883), German inventors and engineers, contributed to the invention and development of electric machines, as well as to perfecting electrical science. The modern unit of conductance is named after them.

Heinrich Rudolph Hertz (1857–1894), German scientist and experimenter, discovered the nature of electromagnetic waves and published his findings in 1888. His name is associated with the unit of frequency.

Questions and Tasks to Chapter 2

1. Whose name is linked with the invention of electricity transmission over long distances?
2. Who and when was built the first overhead power transmission line?
3. What are the names of 2 scientists who have made the greatest contribution to the theoretical and experimental investigation of a rotating magnetic field?
4. What is the name of the scientist who made a decisive contribution to the development of three-phase power transmission lines?
5. What event is considered to be the beginning of electrification?
6. What is difference between concepts of power supply today and in the beginning of the XXIst century?
7. How and where did large scale transmission of electricity started?
8. How are the two events—invention of a rotating magnetic field and fast development of electrical power engineering—connected?
9. What inventions and discoveries in the field of electricity do you assign to the most outstanding?
10. Call names of 5–8 scientists who have made the greatest contribution to science of electricity and its practical application.
11. In what periods of electricity and electrical power engineering history were inventions that are more important made?

References

1. Chang SH (ed). Nuclear power plants. InTech; 2012. p. 350.
2. Warne DF. Newness electrical power engineer's handbook. House Elsevier; 2005.

Chapter 3
The Earth's Energy Resources (Reserves, Short Characteristics)

3.1 Traditional Non-renewable Energy Resources

From a large variety of energy resources occurring in nature and mentioned in Chap. 1, we recognize the *main* resources used in large amounts for human needs.

Shares of energy resources of different types in the global generation of primary energy late in the 20th century are shown in Fig. 3.1 and Table 3.1. They present global energy resources for the same period.

The data in Table 3.1 are temporary and are approximate for three reasons: (1) continuous intake of mineral energy resources takes place simultaneously with the discovery of new deposits, (2) the share available for the power of non-renewable and renewable sources of energy increases due to technological progress, and (3) for commercial reasons, not all countries report reliable data on energy reserves.

The energy of the required type is produced and delivered to customers in the process of *energy generation* that involves five stages:

1. Production and concentration of energy resources including extraction and enrichment of a fuel, creation of a hydrostatic head with hydraulic facilities, etc.
2. Delivery of energy resources to power engineering systems.
3. Transformation of primary energy into secondary energy most convenient for distribution and consumption; electric power and heat are most typical.
4. Transmission and distribution of the transformed energy.
5. Primary and transformed energy consumption.

Figure 3.2 shows the energy utilization scheme.

Figure 3.2b shows the approximate (average) global consumption of energy resources. In this case, setting the total energy of consumed primary energy resources at 100%, the useful utilized energy is only 35–40%. The remaining part is lost; moreover, the largest part is lost as heat. To generate electric power, the mechanical energy of wind ("light-blue coal"), tidal energy ("dark-blue coal"),

© Springer International Publishing AG 2018

V.Y. Ushakov, *Electrical Power Engineering*, Green Energy and Technology,

https://doi.org/10.1007/978-3-319-62301-6_3

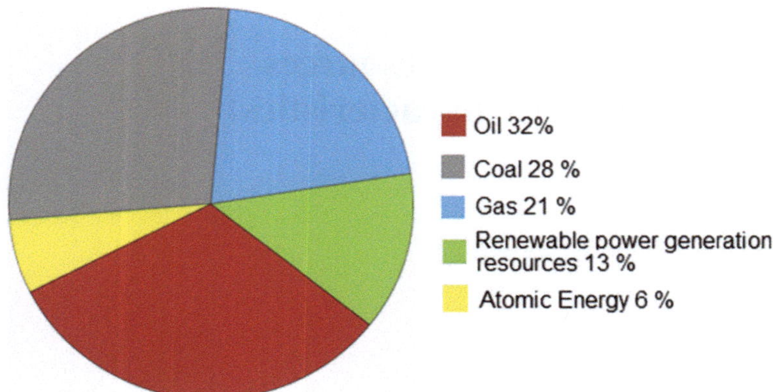

Fig. 3.1 The proportion of various primary energy resources in the world energy consumption

thermal energy of bowels of the Earth (geothermal energy), and solar energy ("yellow coal") (see below) are used in addition to the main energy resources.

The distribution of energy resources over the Earth, countries, and territories inside countries is very uneven. Regions with maximum concentration of energy resources do not coincide with regions of their consumption, as exemplified by oil. More than half the global oil reserves are concentrated in the Middle East, while the level of energy consumption there is less than one fourth of the average global level [1, 2].

Due to the concentration of energy consumption in the most developed countries, we are in the situation in which 30% of population consume 90% of generated energy, and 70% of population consume only 10% of generated energy. The tendency toward an increase in the no uniformity of energy consumption per capita is retained during many decades. Expediency of energy transfer at a certain distance is determined by the *energy content of the energy carrier* defined as the amount of energy per unit mass of the physical body (see Table 3.2).

Among the energy, carriers used at present the highest energy content have uranium and thorium radioactive isotopes. Their energy content reaches 2.22 GWh/kg (8×10^{12} J/kg). Due to a very high-energy content of nuclear fuel, it can be easily transmitted at long distances, since comparatively small amounts of this fuel are required to provide the operation of high-power electrical equipment systems. The power content of fuel averaged over all fuel types is 0.834 kWh/kg (3×10^6 J/kg).

Organic heat by virtue of its specific properties and historical conditions remains the main source of energy consumed by humanity, Fig. 3.1.

Reserves of all fuels that can be extracted from the bowels are limited and according to the data of the World Energy Conference (WEC) are 28.3 million TWh or 3480 billion ton of standard fuel.

Table 3.1 World energy resources, billion ton of standard fuel

Energy sources	Energy resources	
	Theoretical	Technical
I. Non-renewable		
1. Energy of combustible minerals:		
Coal	17,900	637
Oil	1290	179
Gas	398	89.6
Peat	500	100
2. Atomic energy	67,200	1340
II. Renewable		
1. Solar energy at:		
Upper atmospheric boundary	197,000	6140
Earth's surface	81,700	2460
Dry land surface	28,400	3690
Surface of the World Ocean	53,300	
2. Energy of wind	21,300	22
3. Heat of bowels (to depths as great as 10 km):		
Geothermal heat flux that reaches the Earth's surface	3.69	0.35
Hydrothermal resources	1350	147
Underground geothermal resources	36,900	3070
4. Energy of the World Ocean:		
Salinity gradient	43,000	430
Thermal (temperature) gradient	12.3	0.61
Currents	8.6	0.12
Tides	3.2	0.86
Surf	1	0.02
Wind-driven sea waves	2.7	0.1
5. Combustible power resources (biomass) of:		
Dry land	44.2	4.9
World Ocean	23.3	1.84
Organic wastes	2.5	1.23
6. Water power	4.1	1.84

As is well known, the standard fuel unit is taken to mean the amount of fuel the combustion of 1 kg of which yields 29.3 MJ (8.12 kWh) of heat. This unit is used to compare different fuel types.

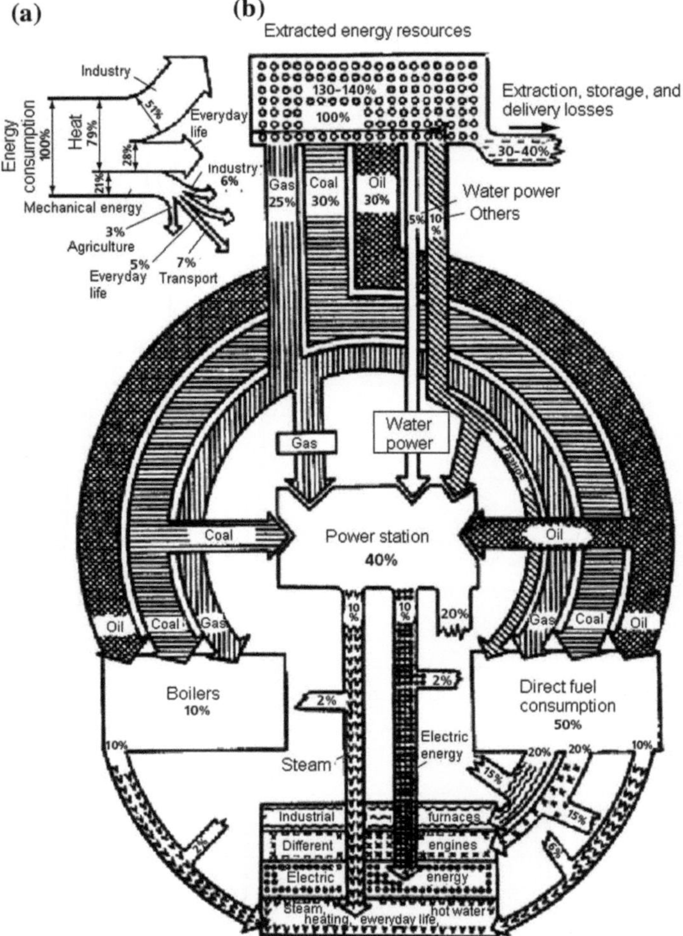

Fig. 3.2 Energy utilization: **a** distribution of mechanical energy and heat delivered to customers and **b** total distribution of energy resources

3.1.1 Coal

The estimated global geological reserves of coal are 61–114 million TWh (7500–14,000 billion ton of standard fuel), from which 24.4 million TWh (3000 billion ton of standard fuel) are proven reserves. Russia and the USA have the greatest amount of proven reserves. Germany, the People's Republic of China, and some other countries have considerable amounts of proven reserves, Table 3.3.

The volumes of coal production in a number of countries are shown in Table 3.4, and the volumes of electricity produced from coal—in Table 3.5

Table 3.2 Characteristics of energy resources

Energy resources: type and sort	Calorific value	Explored reserves	Annual output, generation
1	2	3	4
Gas: Natural Liquefied Casing-head (oil) Casing-head (coal)	8400 kcal/1000 m^3 7600 kcal/1000 m^3 8000 kcal/1000 m^3 6700 kcal/1000 m^3	50 trillion m^3	570 billion m^3
Oil: Crude Fuel Diesel fuel	9600 kcal/kg 10,000 kcal/kg 11,000 kcal/kg		306 million ton 60 million ton
Coal: Brown Black Anthracitic	3400 kcal/kg 5600 kcal/kg 6000 kcal/kg	200 billion ton	200 billion ton 120 billion ton 60 billion ton 20 billion ton
Others: Wood Peat Wood by-products	4000 kcal/kg 2900 kcal/kg 2000 kcal/kg	80 billion m^3 200 billion ton 16 billion m^3	5.2 million ton 0.6 million ton 0.3 million ton
Electric power: CPS DHS HPS APS DPS Unconventional	320 g st. fuel/kWh		850 billion kWh 1228 billion kWh 242 billion kWh 391 billion kWh 24 billion kWh 13 billion kWh
Thermal energy from centralized sources: Electric power stations Boilers	7000 kcal/kg st. fuel		1520 thousand Gkal 596.4 thousand Gkal 526.4 thousand Gkal 60.2 thousand Gkal 319.2 thousand Gkal
Energy saving: Organizational Technological Investment	Amount of the substituted energy resource	400 million ton	50–60 million ton 6–8 million ton 10–18 million ton 25–35 million ton

Quite a lot of coal consumed electricity of Australia, Republic of South Africa, Poland, Rumania and some others countries.

Due to large volumes of gas extraction and relatively low its price the so-called "gas pause"—the rapid increase in the share of gas in the generation of electricity and thermal energy—came in the 70–80s in the Russian energy sector in the 70–80s.

Table 3.3 Coal reserves in the 15 leading countries

Country	Coal	Lignite	Total	%
USA	111,338	135,305	238,308	28.9
Russia	49,088	107,922	157,010	19.0
China	62,200	52,300	114,500	13.9
India	90,085	2360	92,445	10.2
Australia	38,600	39,900	78,500	8.6
Rep. of South Africa	48,750	0	48,750	5.4
Ukraine	16,274	17,879	34,153	3.8
Kazakhstan	28,151	3128	31,279	3.4
Poland	20,895	14,000	14,000	1.5
Brazil	0	10,113	10,113	1.1
Germany	183	6556	6739	0.7
Colombia	6230	381	6611	0.7
Canada	3471	3107	6578	0.7
Czech Republic	2094	3458	5552	0.6
Indonesia	740	4228	4968	0.5
Total in the world	**478,771**	**430,293**	**909,064**	**100.0**

Table 3.4 Coal extraction in leading countries

Countries	Coal extraction (million tons)	%
The whole word	7695.4	100.0
China	3520.0	49.5
USA	992.8	14.1
India	588.5	5.6
EU	576.1	4.2
Australia	415.5	5.8
Russia	333.5	4.0
Indonesia	324.9	5.1
Rep. of South Africa	255.1	3.6
Germany	188.6	1.1
Poland	139.2	1.4
Kazakhstan	115.9	1.5

This allowed rapid increase in thermal electric power with simultaneous reduction of the environmental impact. Today, its share is 42% (the largest among developed countries).

The medium-term plans envisage a gradual reduction in the fuel balance of the share of gas and, on the contrary, an increase in the share of coal, especially in the

Table 3.5 Use of coal in the energy sector in several countries

	Country	Electricity production (TWh/year)	Share (%)
1	India	460	78
2	China	1150	75
3	USA	2200	>50
4	Germany	300	>50
5	Great Britain	135	35
6	Japan	240	<25
7	Russia	170	<20

energy sector of Western Siberia, where one of the most powerful coal basins—Kuzbass is disposed.

3.1.2 Oil

In the modern global economy, the mechanisms of investment and support of its save development are based on oil. Oil is a universal product, which has no analogs. The price of oil is not determined by its production cost, and the factors, which lie outside of the law of supply and demand (the law of the free market), is forecast zone. No one analyst was able to predict price jumps in the oil market. The price of oil is always the result of political (power) compromise. Therefore, the history of oil is the history of wars, coups, and revolutions, that is, the history of our world. The basis of real (not camouflaged) conflict is the struggle for the backbone (political) of life of the world economy—for oil.

Recognition of petroleum as a product—the beginning of trading—goes back to the middle of the XVIst century. In Russia, the first oil was discovered under the river Ukhta in 1721. In 1823 in the North Caucasus kerosene was obtained for the first time using a distillation cube of oil. The first oil company "Pennsylvania Rock Oil Co." was founded in the United States in 1854. In 1859 the company first oil was extracted from the well, and not from the draw-well as before.

Estimation of the global oil reserves is of special interest now. This is caused by a rapid growth of oil consumption and by the fact that in many countries (Japan, Sweden, etc.) oil has substituted for coal in the electric power generation, though this process slows down.

The estimated global geological reserves of oil are 200 billion ton from which 53 billion ton are proven reserves. More than half the proven reserves of oil are concentrated in the Middle East countries. Relatively small reserves of oil are concentrated in Western European countries with highly developed productive forces. The global oil and gas reserves explored late in the 20th century, according to the international "Oil and Gas Journal," are given in Table 3.6.

The estimated proved reserves of oil are dynamic in character. Their value changes after exploration of new deposits, Fig. 3.3 and Table 3.7.

Table 3.6 Global explored oil and gas reserves

Country, region	Oil reserves (billion tons)	Gas reserves (trillion m^3)
Russia	6.7	48.0
USA	2.9	4.6
Canada	0.7	1.8
Mexico	3.4	0.8
Western Europe	2.5	4.3
Africa	7.0	7.3
Countries of Middle East	56.0	42.0
Asian-Pacific region	5.9	9.5
Eastern Europe and countries of the former USSR (without Russia)	1.3	6.0
South and Central America	12.0	6.3
Total	139.2	144.0

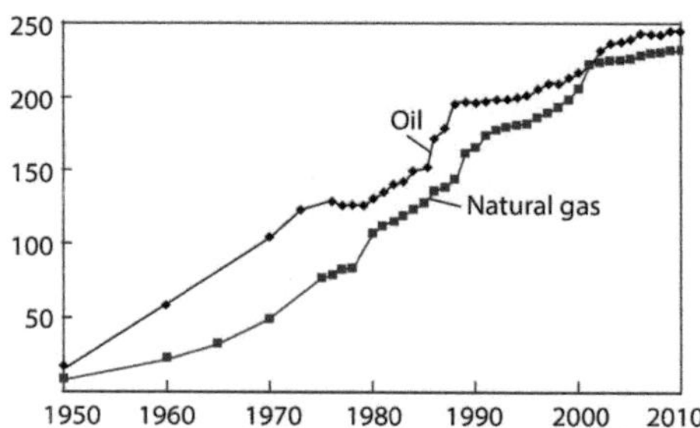

Fig. 3.3 The dynamics of the world oil and gas reserves, million tons of standard fuel (s.f.)

Table 3.7 Production of electricity from gas in 4 countries with the most powerful energy economy

Country	Electricity production (TWh/year)	Proportion of electricity, produced from gas (%)
USA	650	17
Russia	380	42
Japan	260	25
Great Britain	145	40

Extensive geological exploration, as a rule, increases proved reserves of oil and extracted volumes. All estimated reserves from the literature data are preliminary and characterize only the order of the parameters [3]. Figure 3.4 shows the

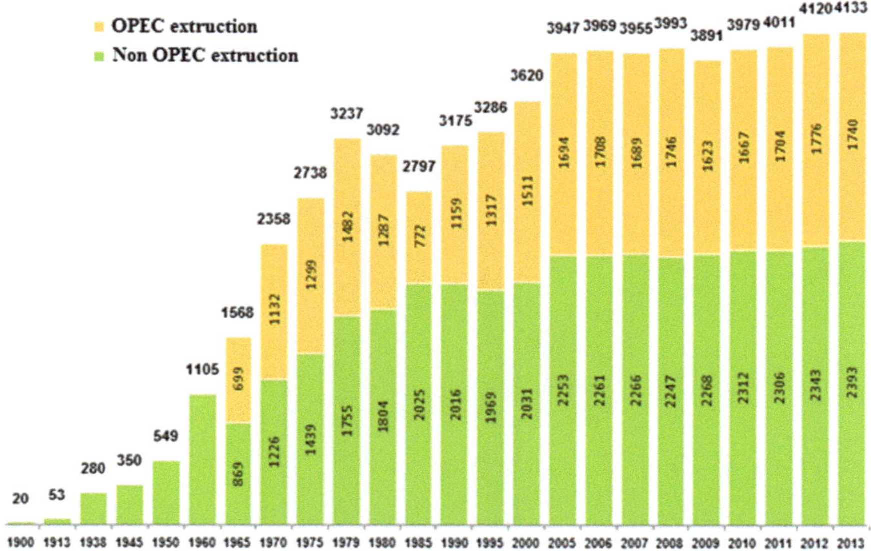

Fig. 3.4 The dynamics of the world oil extraction, million ton

dynamics of the world oil extraction. A very rapid growth of oil consumption is determined primarily by four factors:

1. Development of transport of all kinds and first, cars and aircrafts for which liquid fuel is irreplaceable in the meanwhile.
2. Higher efficiency of extraction, transmission, and utilization compared to solid fuel.
3. The tendency toward utilization of natural energy resources in the shortest possible time and with minimum expenses.
4. The tendency toward earning maximum profit through exploitation of oil deposits.

An important trend in the oil industry is the development of its hard recoverable reserves located in layers with the following features: (a) low permeability, (b) with high water content, (c) with low oil saturation, and (d) with small thickness.

The fact that oil resources are far removed from places of their consumption or centers of location of productive forces resulted in a rapid growth of oil transport means, in particular, in the creation of pipelines of large diameters (exceeding 1 m) and large oil ships.

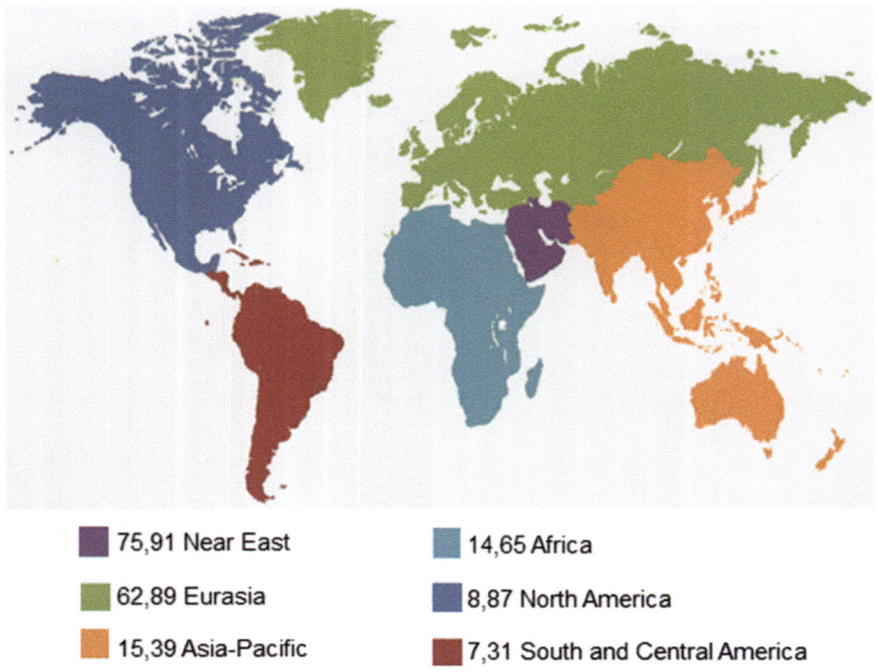

Fig. 3.5 Proven world reserves of natural gas in the 2008, trillion m^3

3.1.3 Natural Gas

The estimated global geological reserves of natural gas are 140–160 trillion m^3, from which ~50 trillion m^3 is a share of Russia and 42 trillion m^3 is a share of Middle East countries. The distribution of gas reserves over countries and continents is presented in Fig. 3.5.

Figure 3.6 shows the world's largest gas fields.

These numerical estimates, as in the case of oil, are approximate. They are changing in the process of exploration of new deposits.

The importance of gas for power is demonstrated in Table 3.7.

3.1.4 Nuclear Energy

The concentration of uranium in the Earth's crust is 0.003%. The amount of uranium in the lithosphere with thickness of 20 km is estimated to be 1.3×10^{14} tons.

The estimated heat content of geological uranium reserves all over the world exceeds by 320 times the heat content of the global reserves of mineral oil.

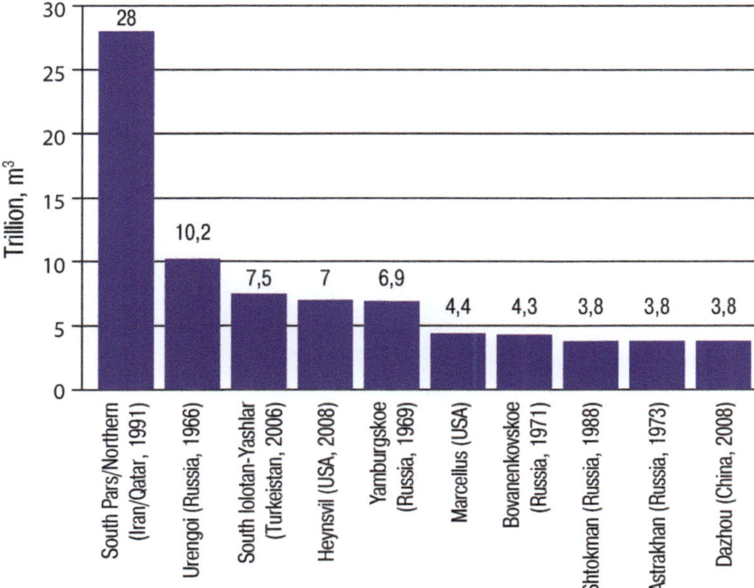

Fig. 3.6 The world's largest gas fields

Table 3.8 Distribution of proven uranium reserves among different countries

Country	Share of world reserves (%)
Australia	37
Kazakhstan	18
Canada	16
South Africa	7
Namibia	6
Uzbekistan	6
USA	4
Russia	3
Brazil	2
Nigeria	1

However, the geological estimate is of limited practical usefulness, because the efficiency of uranium extraction depends primarily on its concentration in reserves. With good quality of natural uranium, ore concentration there is only 0.712%, that is, 7 kg per ton of ore. Some of the richest uranium ore (uranium content of 0.12%) is located in the territory of Canada.

According to the International Atomic Energy Agency (IAEA), the total amount of uranium that can be produced at a sufficiently low cost (less than 22 dollars per 1 kg) is 10 million ton. It will be possible to produce approximately 10 times larger

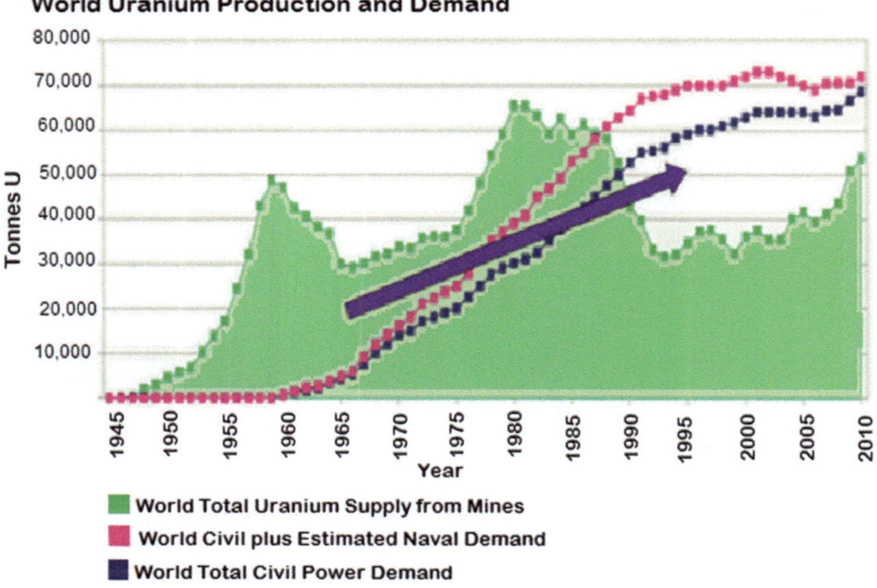

Fig. 3.7 Production and consumption of uranium in the world during the XXth and the beginning of the XXIst century

amount of uranium at 2–3 times greater cost. It is believed that the global power resources will be doubled in the case of complete utilization of nuclear energy.

Just like other energy resources, distribution of uranium ore reserves across countries and continents is very uneven, Table 3.8.

In general, the situation with the nuclear fuel in the world can be characterized by the following facts:

- The annual uranium fuel consumption of about 64 thousand tons (Fig. 3.7).
- Annual production—about 55 thousand tons. The deficit is covered by stocks (armories and civil).
- Term of exhaustion—at least 100 years while maintaining the existing technologies of nuclear fuel—nuclear fuel Open NFC based on reactors with thermal neutrons (RTN).
- In such reactors, U-235 content in natural uranium of only 0.72% is used, and the main component is U-238 (99.28%), the probability of its fission in RTN is very low.
- Projected world consumption of uranium by 2035 will amount to 98–136 thousand tons (large spread is due to a number of policy uncertainty regarding the future of AE).
- In 2010, more than 2 billion dollars were spent for exploration work in the world.

- Urgent is saving of uranium due to: (a) a more complete extraction of U-235 from the ore, (b) improvement of the technology of nuclear fuel cycle, including transition to breeders (fast neutrons reactors—FNR).

However, it is well to bear in mind that in addition to the apparent advantages, atomic power engineering has three severe disadvantages: (a) a small fraction of fuel "burning away," (b) a small but as demonstrated by the history, finite probability of emergency, and (c) problems of burial of radioactive wastes.

Since the chain reaction in modern nuclear reactors proceeds in uranium rods or in graphite and uranium mixture only with the participation of the U-235 isotope rather than in the entire uranium mass and the relative content of this isotope in the basic element U-238 is only 0.7%, methods of more efficient utilization of uranium have been searched already from the very beginning of the nuclear era. The emergency at the Chernobyl NPP on April 26, 1986 took lives of tens and health of tens of thousand people. It dispelled the myth that NPP are safe. Reactions of simple people were identical: they all were in fear of this source of electricity and heat. Enormous damage caused to the population and the nature of the accident at the NPP "Fokushima-1" in Japan in 2011.

Experts have made the only possible rightful conclusion to modify the design of the reactor and all its systems to ensure actual rather than mythical safety of its operation and to avoid building nuclear power plants in areas of high geological activity (threat of earthquakes and tsunamis).

3.2 Backup Fuel (Subsidiary Mineral Fuel)

Experts have long shown interest in the mineral fuel that is not relevant to a classical energetic triad (oil, natural gas, and coal) and nuclear fuel, its share in the energy balance makes about 80%. These fuels (backup fuel or subsidery mineral fuel) include: (1) slate coal, (2) bituminous sandstone, (3) gas—hydrate, (4) associated petroleum gas, and (5) mine methane.

With the improvement of production and utilization in the course of scientific and technological revolution, their share in the energy balance will increase. However, because of the great technological difficulties in the near future they will not be an alternative to "fuel triad" (oil, natural gas, and coal) and uranium, but only helpers. Exclusion can make slate coal as a source nicknamed shale gas.

3.2.1 Slate Coal

The significant part of the world stock of **slate coal** containing such combustion agent as *kerogen* is concentrated in the USA (Colorado, Utah, and Wyoming) and counts about 400 billion ton that equals to 320–630 mlrd tons of liquid hydro-carbon. Slate coal (SC) resources of Russia make 45 billion tons.

SC production in the whole world has been unprofitable recently since it required mine method of extraction, the more so that SC caloric content is inferior to mineral carbon. One ton of SC is needed to get from 0.5 to 2 barrels of combustible liquid that resembles oil. However, the remains of waste material make 770 kg, in addition, they need a place to be stored. The amount of waste material exceeds the amount of initial slate coal; moreover, there are mercury, cadmium and lead emissions; underground waters get polluted by insoluble impurities.

The situation had radically changed when the technology of horizontal well drilling was adapted. The technology made it possible to extract fluid components of these deposits. Experience of technology use in the USA has shown that it increases the wells debit by a factor of at 2.5–3.

In the gas sector in general and in the development of shale deposits in 2009 started a "new energy revolution"—a large-scale extraction of shale gas began.

In 2009 the US produced 745.3 billion m^3 of gas and occupied the first place in the world. This amounted to 40% of shale and coal bed methane. In 2010, the US produced 51 billion m^3 of shale gas. By the beginning of 2010 in the US market over-supply of gas had been formed. So great that the terminals built in the United States on imports of liquefied natural gas remained inactive, then they started renovations to export gas.

In the US, shale gas proven reserves amount to 24 trillion m^3, or more than 10% of the world reserves(at the moment, 3.6 trillion m^3 are technically recoverable). Large deposits of shale gas were found in a number of European countries, in particular, in Austria, England, Hungary, Germany, Poland, Sweden, Ukraine, Romania, as well as in Australia, China, India, Canada. The distribution of shale gas reserves by regions of the world is shown in Fig. 3.8.

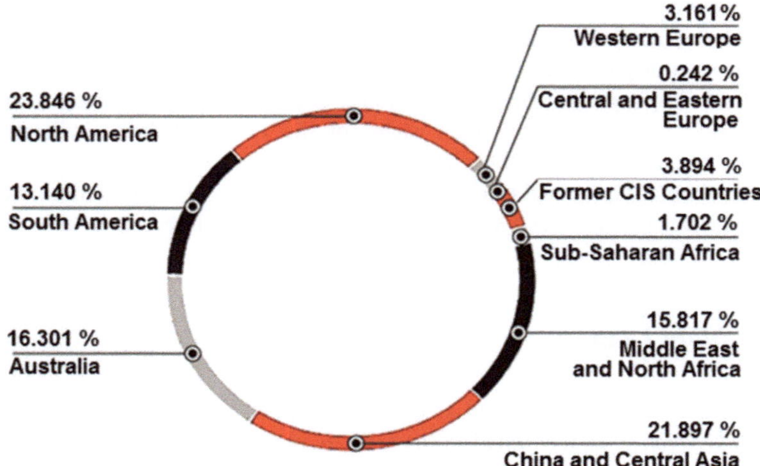

Fig. 3.8 Approximate reserves of shale gas in the world

The IEA predicts that by 2030 unconventional gas in Europe will amount to 15 billion m^3 per year. According to the most optimistic forecasts, the current production in Europe will not exceed 40 billion m^3 per year by 2030. China plans in 2020 to reach the level of production in the range from 60 to 100 billion m^3 of shale gas annually. The wide distribution of shale gas on the planet is its major advantage over conventional one.

The scale and pace of development of work on shale gas will depend on the success in solving number of problems:

- the extraction of shale gas is effective only by horizontal wells, which cost 4 times higher than the vertical; the problem is aggravated by the higher hardness shale compared to rocks in the fields of conventional gas;
- shale gas lies in a small isolated "pockets"; to get it in the appropriate volumes, it is necessary to drill a large number of wells, each of which has a small production rate; it is unacceptable for countries and regions with high density of population (in particular, for the overcrowded Western Europe, in dire need of gas, but reverently relating to the environment, including the not yet built-up land); well service requires a large number of roads and pipelines to deliver a huge need for the volume of water;
- for efficient extraction of gas from shale, fracturing should be implemented— water with special chemicals should be pumped into the well under high pressure, tearing walls between the pores, and with the sand that riving these gaps; these chemicals can get into the aquifers and adversely affect the ground water—in fact, the last reserve of drinking water in the world;
- the life of horizontal wells is estimated to be 5–12 years, whereas conventional wells are in operation for 30–40 years;
- according to independent experts, the cost of production 1000 m^3 of shale gas in 2009–2010 was \$212–283, and the volume of production from a single well— about 50 million m^3 (companies extracting shale gas, probably for promotional purposes referred to other values—\$150, and 180–200 million m^3, respectively);
- in the process of horizontal drilling, fracturing and completion of wells in the atmosphere gets that amount of methane that it negates all the benefits of shale gas relative to coal.

From 2010 to the summer of 2014, many members of the slate industry switched from production of shale gas to the shale oil (Fig. 3.9). In 2014 the United States produced about 0.4 million barrels of shale oil per day.

Around the shale gas hot economic and political debates on the global level and in a number of countries and, above all, in the United States were opened. In 2015 because of the collapse of prices for conventional oil a number of US companies, producing shale oil became unprofitable and practically ceased operations.

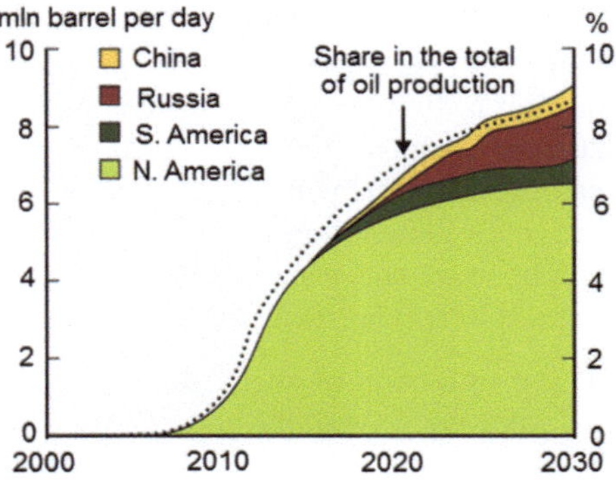

Fig. 3.9 Extraction of shale oil in several countries

3.2.2   Bituminous Sandstone

Bituminous sandstone (BS) is a mixture of sand, clay, and oil asphalt. One ton of BS, containing 14% of bitumen, needed to get 100 L of liquid carbohydrates. The significant part of the world stock of BS is concentrated in the USA, Canada, and Venezuela. In some developed countries production and processing of BS has started aiming at extraction from it non-standard oil. However, this type of oil is being produced now only in Canada where BS contains bitumen in large proportions. According to some predictions, in 2010 BS production will increase from today's 1 million barrels per day to 1.3 million barrels, and by 2015 will exceed 2.7 million barrels per day. According to some US experts, BS processing is profitable only if the oil price is no less than 100–200 dollars for a barrel.

High working cost interferes with the intense processing of bituminous sandstones and slate coals. The estimated cost of 1 m^3 bitumen production from sandstones is 220–314 dollars, from slate coals is 340–350 dollars. Thus, non-standard oil production is near at hand.

3.2.3   Gas Hydrate

Gas hydrate (GH) is a peculiar type of gas—a crystalline molecule water cell with the methane molecule inside (Fig. 3.10). The cells form a compact crystalline grid that resembles ice (Figs. 3.10 and 3.11).

Escaping methane from deep under ground is able to create GH only with water combination, particular environmental conditions, low temperatures and high pressures. That explains the reason why GH can be found only in seas, oceans and permafrost areas. Gas hydrates can be formed in bottom sediments with the pressure of 25 atmospheres and temperature of about 0 °C. In case of higher temperature, a higher pressure is required in order to form GH. Gas-hydrates can be usually found in forms of layers in oceans and seas at depth from 300 to 1200 m. Rarely they can

Fig. 3.10 The combustion
gas hydrate

be found close to the sea bottom (at a depth of several meters from its surface). On a Messoyakhskiy natural (ordinary) gas field in Western Siberia (on the left bank of the river Yenisei) GH appear even on the surface.

For the first time GH of marine occurrence have been detected in the mid-seventies by the Canadian fisherman, and GH, laid out under ever-frozen layers, were found by Academician A. Trofimuk et al. in 1975.

Special activity in prospecting GH and developing economically and ecologically effective technology of methane withdrawal from them is exhibited by the countries experiencing deficiency of power resources—Japan, South Korea, Germany, the USA, Canada, China, etc. For today, more than 200 fields are opened in various parts of the globe, Fig. 3.12. Canada has detected vast gas hydrates (Mallic) in river Mackenzie plain with stocks of gas of 110 billion m^3, Japan lays out gas stocks in gas-hydrates on Pacific shelf (field Hankai) in bulk of 4–20 billion m^3.

Thanks to the studies held in the 80–90s, gas stocks of GH in continental and shelf parts of Russia are valued by 100–1000 billion m^3. Reservoirs of GH are detected also in the Black sea and Lake Baikal.

Owing to a poor level of scrutiny of genesis of gas hydrates and small volume of search and trial works, world's reserves of gas-hydrates on different sources lay in

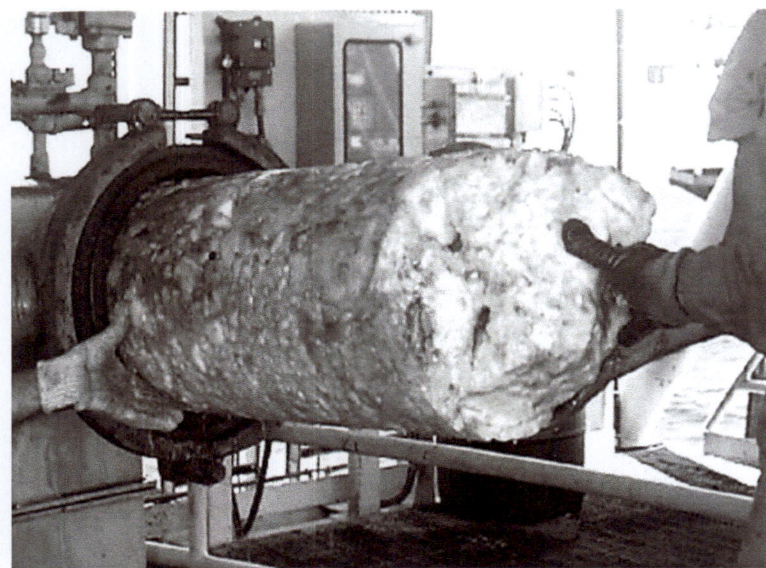

Fig. 3.11 Gas hydrate cork in the pipeline

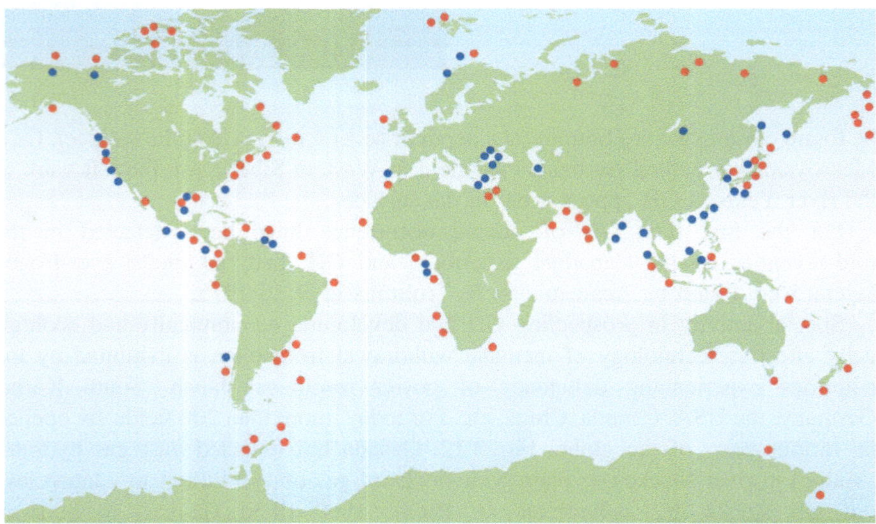

Fig. 3.12 Location of deposits of GH (*Blue dots* the places where the presence of GH exploration drilling confirmed, *red dots* confirmed indirectly by geophysical exploration)

extremely wide limits—from several hundred to several thousand billion cubic meters. It is considered that GH contains raw hydrocarbons 10 times more, than in all oil and natural gas fields. GH are unique yet not developed and sufficiently perspective source of gas distillate (from 1 m^3 of gas hydrates it is possible to get 300 m^3 of methane). If the problems concerning extraction of methane from GH are solved successfully, the shape of the world market of power resources can be essentially inflected, and the problem of "hunger for energy" will be removed for uncertain period.

In the USA in 2000, the law has been passed providing allocation within five years 47.5 million dollars for works on investigation and gas production from gas hydrates. Even more active in this field is Japan which intended to start industrial extraction of methane from gas hydrates within the next decade.

In Russia, according to the offer of Rosnedra, two centers with financing of 20 million rubles for research of GH in the Black Sea and the Sea of Okhotsk will be allocated until 2010.

Difficulties of extraction of methane from GH consist in that at their lifting on the surface they quickly decay into methane and ordinary ice. Thus, methane disappears, aggravating the "green-house effect". In this connection, it is necessary to master technology of extraction of gas directly in depth by means of deep-water extracting platforms, from where gas will be pumped by pipes in special tankers.

3.2.4 *Associated Petroleum Gas*

For today more accessible and competitive technology of use of "backup" fuel is oil recovery and subsequent recycling of **associated petroleum gas** *(APG)* which represents a mixture of methane and a wide spread of a light hydrocarbons (ethane, propane, butane, isobutene, etc.), and also non carbon components (nitrogen, carbon dioxide, hydrogen sulphide, helium, and argon). By the geological characteristics, the associated petroleum gas caps and the gases dissolved in oil are distinguished. The first are evolved from the oil wells, the second—from reservoir oil in the process of its separation. APG is presented far not in all oil deposits, and where it is, volumes fluctuate from units to several thousand cubic meters per one ton of the taken oil.

APG represents valuable power raw materials thanks to its high caloric content and environmental friendliness, and to the convenience of supply of oil fields with the energy from means of small power stations. As the raw materials for oil-and-gas chemistry, APG allows to receive a net gas similar to natural one, a casing-head gasoline, liquid gas for everyday needs, and light hydrocarbons that by combustion can be recycled into rubber, plastics, and components of high-octane petrol. Russian mass media are telling very much about APS combustion in flares (torches above oil fields) as symbols of mismanagement and barbarian attitude to nature.

Combustion of APG is admitted as a problem number two in the activity of Russian fuel-energetic complex (small percentage of completely recycled oil for

Fig. 3.13 Gas torches on the offshore and onshore Russian oil fields

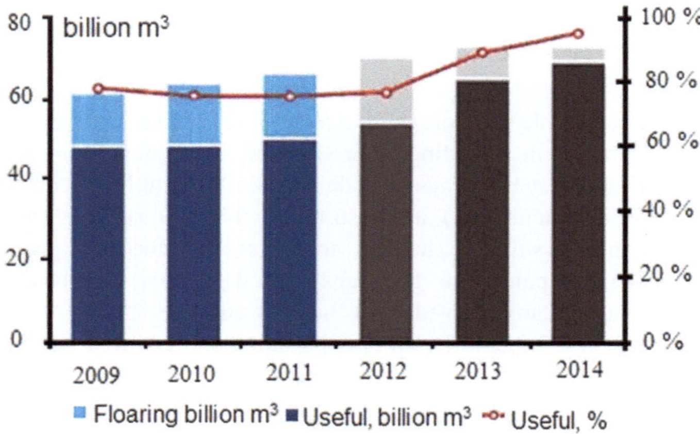

Fig. 3.14 The volume of associated petroleum gas that was burned and beneficial by used in Russia

receiving products with high additional cost is considered to be the problem number one), Fig. 3.13.

Nowadays the problem of useful APG is successfully solved in many countries, including Russia, Fig. 3.14.

The following factors were obstacles for beneficial use of APG for many years:

- It is difficult and expensive to collect APG. That is why it turns much easier for oil companies to pay little tax for atmosphere pollution then to invest money into APG recycling. The APG price regulated by the government is about 400 rubles for one hundred m^3. This makes its sale to native home consumers not profitable and tax for burning gas is little (about 40,000 rubles) and can be taken only once a year;
- in the period of prosperous growth of oil extraction, building of infrastructure for APS utilization was not in time with high rate of extracting main raw materials;
- Law about "associated petroleum gas", which allows making order in the question of APG, was accepted only in 2009.

Oil companies have to elaborate the new investment projects in order to change this situation, and the state has to provide the favorable conditions for their realization.

The projects have to be economically profitable or, as minimum, non-unprofitable in the first stage for oil companies.

Producer of APG ought to have access to gas pipeline system belonging to the company "Gasprom". This problem is especially difficult for solving in winter, when gas pipeline is overloaded. For this purpose, for example, the "Holland experience" can be used that is, regulation of standard gas extraction so that to provide APS with admittance to gas pipeline system.

The increase of power of the plants that prepare APG for consumer supply (removing petrol from the APG) have to be coordinated with the volume of APG extraction and gas pipeline possibility to take the prepared APG.

In the second half of 2007, the Ministry of Natural Recourses brought to the government of the Russian Federation the project on measures aimed is to rise effectiveness of using APG. The document considered the introduction of 95 percent norm of recycled APG to 2011. From 1 January 2011, a progressive scale of payment for pollution caused by APG has been introduced. The measures of encouragement have also been foreseen for its using: providing a priority access to the United National Energy system of providers of electricity produced on power stations from APG, tax introductions, and customs privileges for enterprises implementing projects on rational use of APG. At the last few years under pressure of federal and local administrations, as well as from economic considerations, oilmen have begun to change its attitude to associated petroleum gas.

3.2.5 Mine Methane (Coal Methane)

In the nearest future, **mine methane (MM)** can become an observable gain to gas part of energy balance, raise efficiency of small and hydrogen power engineering, provide the transport with the cheap and ecologically clean fuel. Emissions of MM make 10% of all anthropogenic methane emissions. World stock of mine methane

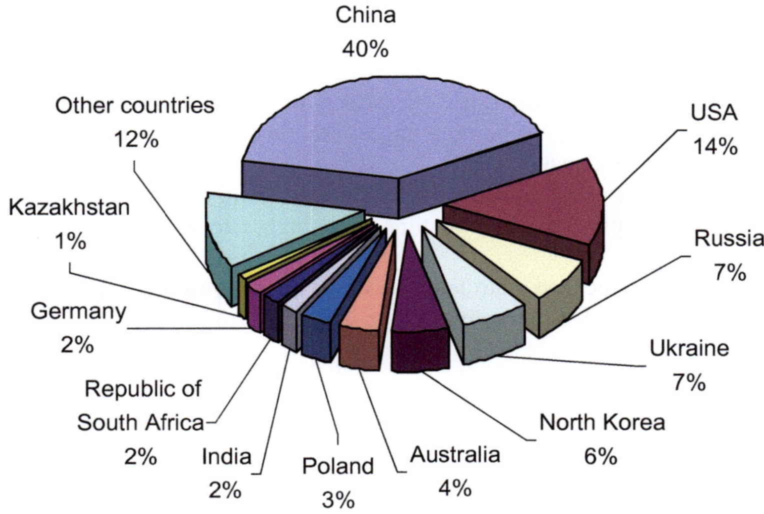

Fig. 3.15 Structure of world's MM emission

makes about 260 billion m^3, and it is they are distributed over the countries as shown in Fig. 3.15.

Only Russian enterprises of coal branch annually extract from coal layers about 2 billion m^3 of MM.

The mine gas is divided into two essentially different classes: (1) ventilating gas with concentration of methane 0.5–2.0% and (2) methane at decontamination of coal layers with concentration of about 30–90%. In the hole volume of emissions, the basic contribution belongs to ventilating emissions—to 85%.

Accordingly, extraction of MM from coal deposits is carried out by two schemes:

(a) Extraction of MM as associated petroleum gas at coal mining,
(b) Extraction of MM from wells specially drilled in a coal file as an independent kind of production.

The first scheme is aimed at increasing safety of miners' work. The second scheme provides preliminary and fuller extraction of MM and essentially reduces probability of explosion at coal mining. Duration of work of wells does not concede to service life of wells on natural (standard) gas deposits. This scheme has been working for a long time in the USA, Canada, Great Britain, Australia, China, Germany, and some other countries. Extraction of MM in the USA exceeds 5% of gas consumption level in the country that makes about 35–40 billion m^3. In these countries MM is supplied into the main gas pipelines, it is used for joint burning with coal in boiler-houses, and also in internal combustion engines and gas turbines.

Extraction of methane from coal deposits has a number of advantages in comparison with other ways of its extraction:

- Majority of coal deposits is located in already habitable areas with developed infrastructure, industrial base and manpower resources;
- basically coal layers are located at rather small depths (to 1000 m);
- Volume of prospecting drilling has essentially reduced, since already known contours of a coal file at the same time are contours of a gas deposit in depth of coal layers.

However, there are also serious obstacles for large-scale extraction of MM. The main is a difficult technology of its extraction in comparison with extraction of standard (natural) gas. The coal layer gives methane at intensive mechanical influences as it occurs in mine or on an open career at destruction of a layer by a combine or an excavator, or by a hydraulic shock, or an explosive by means of wells specially drilled for this purpose.

In Russia, the basic emissions of methane occur at coal extraction in Kuzbas where at the general stocks of methane in coal layers about 13 billion m^3 its annual emissions reach 2 billion m^3. At the same time, to meet internal needs, Kemerovo area imports up to 1.5 billion m^3 methane a year. For a heat supply of the enterprises and municipal service, more than 2 thousand boiler-houses operated on coal fuel.

Management of the Kemerovo area intends to develop accelerated rate technology of recycling MM. It will not only expand resource, but will also give considerable social and economic effect in the form of increase of safety of miners work (approximately twice as much), gasification of the region, and provision of new workplaces.

In the USA the stringent laws oblige mine owners to provide obligatory decontamination of coal layers before their extraction. That is the main reason that explains why miners in the USA perish seldom. In Russia and many other coal-producing countries one perish of a miner annually happens per 1 million tons of coal extracted. It is also necessary to bear in mind that methane, which thrown into the atmosphere, surpasses carbon dioxide by 20 times as "green house" gas.

Russian government has approved the project of building in the Kemerovo region a national industrial park, called to find modern decisions concerning safety support in coalmines. Firstly, the project deals with the total elimination of "gas-killer", i.e., methane (with its subsequent recycling).

3.2.6 Syngas

Syngas (synthetic gas) is produced by gasification of solid or liquid hydrocarbons according to one of the two technologies: (1) unremitting gasification and (2) simultaneous production of syngas, coke, and other products.

The highest prosperity of coal gasification reached in the first quarter of the XXth century. In 1925, about 12 thousand plants in the USA were processing about 25 million tons coal per year into gas. In the 50 s, the USSR produced about 35 billion m^3 of gas per year from coal. The intensive development of oil and natural gas extraction has led to a loss of interest in this technology. However, since the date of exhaustion of oil and gas is substantially closer than of coal, in recent years, interest in this technology has been increased.

Gasification technology is characterized by large "unpretentious" and versatility:

(a) it can be any gasify solid fuels—from the very young peat and lignite to bituminous coal and anthracite, regardless of their chemical composition, the composition of the fly portion of sulfur impurities, particle size, moisture content, and other properties;

(b) Flammable gases can be produced in a wide range of composition and properties: pure hydrogen, carbon monoxide, methane, and mixtures thereof in various proportions by means of gasification.

Components of $(CO + H_2)$ are used as a raw material for a wide variety of chemical synthesis processes, which have already been developed by industry: the production of ammonia, methanol, dimethyl ether (as a promising alternative to diesel fuel oil) and the generator gas that can be used in power plants.

Hydrogen is extracted (20–25%) can be used as a motor fuel, raw materials for electrochemical elements, in biotechnology as a reducing agent in metallurgical processes of direct reduction of iron and other metals.

The obstacles to the expansion of the scale of production of syngas are:

- Higher capital intensity,
- The high cost of gas purification from impurities,
- The need for dozens of gasification processes adapted to the characteristics of raw materials and the requirements for the product (gas). The most promising of these are:

  - Techniques (methods) with a combined production of several products,
  - In-cycle gasification by coal combustion in furnaces of power plants and high-power boilers.

In the coalfields with particularly difficult conditions for coal mining, a promising method of gasification is ***underground gasification.*** In the coal seam, it is expedient to create controlled burning hearth to produce a combustible gas. For this purpose, wells are drilled from the surface for the organization of blast and extraction onto the surface of the product—the combustible gas. All process steps are carried out from the surface without the use of underground working.

Today, research in this area is intensively carried out in China, where dozens of plants have been built in recent years, and in Australia, where a large enterprise of that profile has been built. India, North Korea, and Republic Korea show the interest in this research.

Since the middle of the 90 of the XXth century, *the synthesis of methanol and dimethyl ether* technology based on carbon dioxide industrial emissions are being developed (coal thermal power plants, metallurgical, chemical, cement plants, etc.) and electrolytic hydrogen and oxygen is produced by electrolysis of water using renewable energy sources and nuclear power. The author is a Nobel Laureate George Olaf says that up to 90% of all polymers and 100% of motor fuel and oil can be obtained in this way.

In conclusion, the following should be noted: In technology of extraction and transformation in electric and thermal energy, subsidiary ("backup") mineral fuel so far concedes to traditional mineral fuel. This means that the governments of different countries and the management of corresponding companies should render them comprehensive support, at least until they become competitive enough. In the countries-exporters of power generating resources, it seems to be logical to take means for these purposes from super incomes of the oil and gas companies.

Questions and Tasks to Chapter 3

1. What are the main stages of the life cycle of mineral energy resources?
2. What are the figures reflecting very uneven consumption of electricity residents of highly developed and less developed countries?
3. To what extent does the share of electricity produced from coal change, among the countries with the highest volume of production?
4. What is projected deadline for exhaustion stocks of oil, gas and coal?
5. What are the dynamics of explored and produced volumes of oil and gas?
6. Which of the top 5 countries have the most part of the nuclear energy resources?
7. What limits the use of peat in the energy sector?
8. What are the main types of auxiliary fuel?
9. What are the main problems of extraction of the shale oil and gas?
10. What is it gas hydrate? What are the prospects for the energy use of methane hydrate?
11. What is it associated oil gas?
12. What are the most effective ways of utilization of associated oil gas?
13. What does it mean "the utilization of mine methane (coal methane)"?
14. What problems does the use of auxiliary fuel help?
15. Why governments of countries that produce coal stimulate the collection of mine methane?
16. What stimulates oil producers utilize associated petroleum gas?
17. What are differences between extraction of traditional and slate oil?
18. What are differences between the traditional power resources and backup fuel?
19. What are the functions of OPEC?
20. What is the primary attitude of oilmen toward the petroleum gas utilization?

References

1. Economides M, Oligney R. The color of oil (The history, the money and the politics of the world's biggest business). Katy, Texas: Round Oak Publishing Company; 2003.
2. Warne DF. Newness electrical power engineer's handbook. House Elsevier; 2005.
3. Bauman Z. Leben in der Fluechtigen Moderne [in German]. – Frankfurt am Main; 2007.

Chapter 4
Electric Power Production

Over the past century, the electric power industry continues to contribute to the welfare, progress, and technological advances of the human race. The growth of electric energy consumption in the world has been phenomenal. In the USA, for example, electric energy sales have grown to well over 400 times in the period between the end of the XIXth century and the early 1970s. This growth rate was 50 times as much as the growth rate all other energy forms used during the same period. The installed kW capacity per capita in the USA was about 3 kW.

4.1 Choice of the Electric Power Generation Type

The reserves of power resources and their output are determined by the efficiency of their consumption. Modernization of technical equipment permitting to utilize more completely (that is, with a higher efficiency) the primary energy resources implies that a smaller amount of primary resources is required to generate the same energy amount [1–3]. If we estimate the efficiency of utilization of primary resources in terms of energy output, we are forced to point out that they are transformed into electric power with a very low efficiency at stations of different types (Table 4.1).

Moreover, the efficiency of atomic power stations is greatest, and the efficiency of hydroelectric power stations is least. The energy carrier consumption and the efficiency presented in Table 4.1 were reduced to identical powers (1 GW) of power stations generating 24 GWh a day (86.4×10^{12} J).

The efficiency can be calculated as follows. To generate an energy of 120×10^{12} J, 700×10^6 ton of water must flow through turbines of a 1-GW hydroelectric power station. This water mass has an internal energy of 630×10^{26} J. Therefore, the efficiency of water consumption is $\eta = (120 \times 10^{12}/630 \times 10^{26}) \cdot 100\% = 0.19 \times 10^{-12}$. Analogously, the efficiency of TPP and NPP can be estimated. The efficiencies of fuel consumption at these stations are 25–40 and 15–30%, respectively.

V.Y. Ushakov, *Electrical Power Engineering*, Green Energy and Technology, https://doi.org/10.1007/978-3-319-62301-6_4

Table 4.1 Energy carrier consumption and efficiency of different plants

Amount of energy carrier required to generate electric power of 33.4 GWh (120×10^{12} J)	Type of the station/plants (output power of each station is 1 GW)	Efficiency of energy carrier consumption (%)
700×10^6 ton of water	HPS (hydroelectric power stations)	0.19×10^{-12}
6400 ton of coal 4600 m^3 of oil 536,000 m^3 of gas	TPP	0.2×10^{-5}
1.5–2 kg of uranium	NPP	10^{-2}

An increase in the efficiency of energy generation is a very urgent problem. As a whole, it increases continuously. Thus, the electric power generation all over the world has increased by a factor of 3.5 for 15 years (1950–1965). During this period, the production of primary power resources has been only doubled. This is due to the increased efficiency of electric power systems and the increased amount of primary resources used to generate electric power [4–6].

The global electric power generation increased from 11.5 billion kWh in 1990 to 19.4–24.4 billion kWh in 2020, that is, by a factor of 1.7–2.1. During this period, an advanced growth of the electric power generation compared to the consumption of primary power resources was observed for all variants. The tendency toward an increase in the electric power consumption per capita was also typical. This parameter increased from 2.17 thousand kWh per capita in 1990 to 2.4–3.03 thousand kWh per capita in 2020. However, a new tendency was manifested, namely, a decrease in the electric capacitance of the gross domestic product (GDP) (by 8–27% to 2020 relative to the level of 1990.)

Among specific features of the future global electric power engineering will be centralization of energy distribution and a great diversity of generating sources. Its characteristic features are combination of large-sized powerful centralized and relatively small enterprises operating in a common grid, application of hybrid schemes of electric power and heat generation, compatibility of energy and production technologies with complete utilization of wastes and secondary resources, and formation of interconnected power systems.

Problems whose solution depends mainly on the energy industry initiatives are:

1. Low efficiency of energy production
2. Too small scale of subsidiary (auxiliary) mineral fuel use in the energy sector
3. Small share of the distributed power generation in total world electricity
4. Small share of the renewable power generation in the Fuel and Power Complex.

Since mankind has entered into the next 7th environmental crisis (now it has the global character), we have gradually been involved into technical (technological) revolution. For electrical power engineering this means development of new energy sources and new ways of converting them into electrical and thermal energy. The beginning and pace/rate of this revolution depend upon a number of factors. The factors determining the onset and rate of the Global Energy Revolution development are

1. **Economic**—requires investment on a huge scale
2. **Technology**—the inability of quick change of technological structures
3. **Social**—(a) rapid population growth in developing countries and (b) the inability to quickly change the existing style of life
4. **Global** (geographical)—extremely uneven distribution of energy resources over countries and regions.

The leaders of all levels (state, regional, and municipal), solving the problem of energy supply to population of the territory, have a wide choice. This choice is based on the account of two groups of factors.

1. Condition (characteristic) of the region:

   - The level and profile of the economy,
   - The number and density of population,
   - The availability of energy,
   - Geographical and climatic conditions,
   - The degree of depreciation of fixed assets in the energy sector of the region,
   - The possibilities of domestic power engineering companies and the availability of foreign equipment,
   - The opinion of the majority of the population.

2. The current technological and technical level of energy production:

   2.1. By the type of primary energy: (1) "fuel triad", (2) nuclear fuel, (3) subsidiary mineral fuel, and (4) renewable energy.
   2.2. Power generating plants and scheme of their connection to consumers: the network ("big")—an autonomous/distributed ("small").

The ranges of possible capacities of individual units of various types are shown in Fig. 4.1.

4.2 Powerful Power Plant Based on Non-renewable Mineral Energy Resources

Several factors remain important for power plants using mineral energy resources:

- Mineral fuel resources are sufficient, at least until the end of the XXI century,
- The efficiency of thermal power plants continues to increase (from 10% in 1900 to more then 50% in 2020), Fig. 4.2.
- Fuel energy becomes more friendly to the environment,
- The efficiency of the fuel cell is already more than 70%. When bringing their main characteristics (unit cost of installed capacity, reliability, service life) to the

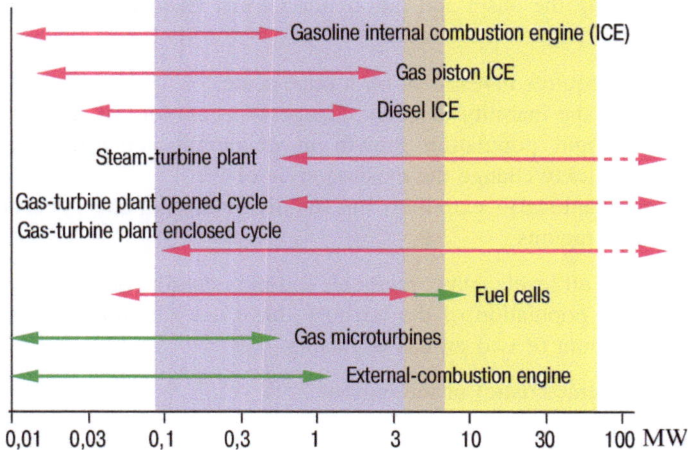

Fig. 4.1 Scale of single power units of various types

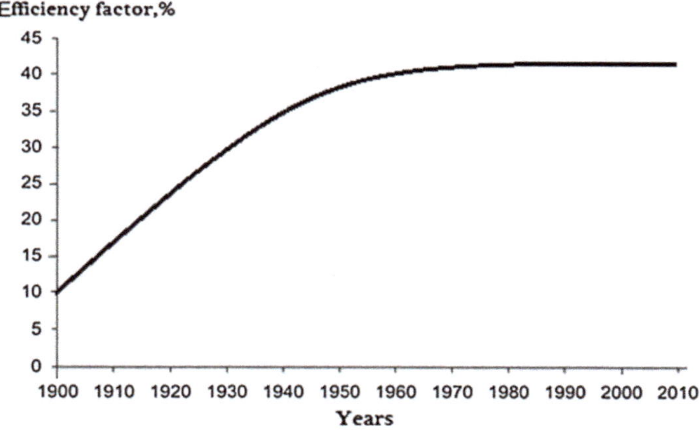

Fig. 4.2 Changing the efficiency factor of thermal power plants throughout the history of their use

level of the stationary energy requirements they may occupy an important place in it,

- Development of alternative forms of power generation (nuclear power plants, large-scale hydroelectric plants, power plants using unconventional renewable resources, etc.) constrained by a number of problems that remain unsolved.

4.2.1   Features of the Use of Coal in the Energy Sector

In modern conditions of strict requirements for environmental protection in the production of electricity and thermal energy, the most complex problems arise in the operation of coal-fired power plants. Besides, they have to compete with other types of power plants. These circumstances and a large proportion of coal-fired power in the total power generating capacity have attracted more attention to them in this textbook.

To troubled branches of the fuel-consuming power industry belong also nuclear power industry and a small (distributed) power industry, which have received considerable attention in this book.

Gas, fuel oil and gas-oil power plants are less problematic in the aspect of environmental safety and energy efficiency; in addition, electricity and heat production technology for them have well been mastered. For these reasons little attention, is paid to them in the textbook.

The "road map" for the power engineering development in many countries foresee considerable reduction of gas volume combusted in furnaces of power plants and boiler rooms and increase in the share of coal.

The main reasons for these changes are:

- Too large share of gas in the energy balance which has been observed in the last few years as an inadmissible imbalance that has appeared for the last 2.0–2.5 decades and nowadays are known as "gas pause". Its origin is caused by the low cost of natural gas in some countries (especially in Russia and some other countries of Middle East) in comparison with other mineral fuels. (In the markets of Europe, the USA, Japan, and South Korea the cost of gas is much higher then that of coal and a little lower then of oil).
- High value of the natural gas as a raw material of chemical industry and hydrogen power engineering.

The coal is becoming more popular as fuel for power plants and boiler rooms due to the action of the following factors:

- world-wide coalfields are very rich,
- according to the most optimistic predictions the rate of nuclear energy in the PE will increase from the present 16.5 to 21–25%,
- rate of hydro energy generation may decrease from 17 to 13.5%,
- Contribution of unconventional renewable power sources will be presented in small amount (the probable percentage will be only 2–3%),
- Oil products will be used as a source of fuel reserve; therefore, their presence in the energy balance will not be high.

To renew the dominant role of coal in the PE, it is necessary to introduce a new set of prompt economic, technological and organization actions:

- change the ratio of gas and coal costs (gas will be 1.6–2 times more expensive than coal),
- increase the coal utilization efficiency, and
- Enhance the coal environmental safety in the PE.

There are several obstacles in this way:

- modern and world-wide known gas-turbine and steam-and-gas production methods of electrical power provide the highest pace of power intensification in the PE (It is of paramount importance due to the increase in energy shortage, "hunger for energy"),
- due to the existing price ratio of natural gas and coal, the largest heat gas-stations are in advance due to two main factors: capital outlay in power plants building and the cost of energy produced,
- the growth of ecological taxes and toughening up the sanctions on the environmental damage force power engineering specialists to improve methods of coal firing and emission purification that considerably increase cost per units of installed power and energy produced.

Taken as a whole, these factors force to put up with the fact that by 2020–2025 the basic signs of "gas pause" in Russia and some other countries will remain, i.e. gas as the primary energy resource will dominate in the energy production. The rate of coal production in the structure of heat power plants should gradually increase from 25 to 36–37% provided that the rate of gas production decreases from 68 to 58%.

The electric and heat energy production by coal-fired power plants is the main cause of environmental pollution:

- emission of (a) carbon, sulfur, and nitric oxides, (b) flue ash, (c) particulate matters,
- coal-volatile matters do damage to the atmosphere when coal is kept on the esplanades of heat power stations,
- the esplanades may cause continuous ecological disturbances such as:

(a) geo-filtration of industrial effluents, (b) erosion and collapse of dam slopes, (c) Spreading of slurry through breaks in dams and flooding of the adjacent territory, (d) dusting ash, etc.

In terms of money, ecological damage in Russia caused by electric energy production coal heat plants makes up 64 cents for 1 kWh, at gas heat plants—2.8 cents, and 0.1 cents at nuclear power plants.

Increase in coal consumption by 2.3 times to 2020 according to "The Power Engineering Strategy of Russia to 2030" seems to be kind of a challenge that can only be accepted by adoption of new technologies guaranteeing environmental protection.

Experience of such developed countries as Germany, Japan, Great Britain, and South Korea proves that the modern level of PE-technologies allows minimizing ecological impact of coal power plants at minimum costs.

In the Russian ecological program a set of activities aimed at technical re-equipment of thermal power plants is foreseen. They, in particular, provide:

- realization of new technologies of coal firing in very powerful coal unit boilers (450 MW and higher) with critical and supercritical flow parameters,
- development of traditional generating sets of coal plants, i.e., higher steam conditions have been used and unit rating has been increased,
- High technology employment of atmospheric, soil, and water emission purification, realization of a set of activities aimed at technical re-equipment of thermal power plants, coal plants in particular.

The main methods of coal firing improvement (to make thermal power plants friendlier to the environment and more effective) include: (1) improvement of the furnace construction and technology of coal firing, (2) increase of the fuel quality.

4.2.2 Improvement of the Furnace Construction and Technology of Coal Firing

Improvement of the furnace construction and technology of coal firing provides:

- Converting the coal-steam power-generating units to units with ultra-critical flow parameters (pressure 24–30 MPa, temperature 580–610 °C, efficiency factor 45–50%),
- putting into operation power-generating units operated by flare coal firing boilers and boilers with circulating boiling bed,
- development of powdered-coal power-generating unit with the maximum steam temperature over 700 °C and pressure 37.5 MPa (these parameters are planned to be achieved by 2015–2020, efficiency factor will exceed 50% and will reach 55% with the steam temperature up to 800 °C),
- low-temperature vortex technology of coal firing,
- Adoption of new technology of coal firing and conversion: coal firing and gasification in melted slag (IGCC-technology), coal firing in boiling bed or in circulating boiling bed, etc.

Combined-cycle plants with closed cycle coal gasification have a number of advantages:

(1) They allow burning down low-grade coal (with content of ashes up to 40–45%),
(2) They also allow reducing the size of gasification reactors and boilers,
(3) simplify a system of fuel preparation and supply,

(4) Capture mineral part of fuel in ash-and-slag melt.

Special programs aimed at ecologically clean coal consumption are approved in Germany, Great Britain, Japan, Netherlands, Russia, and other countries.

There are more then 600 boilers operating on circulating boiling bed, including Europe—275, Japan—28, China—25, Asia on the whole—126, the USA—155.

In Russia the first power-generating unit of this type was built in Novocherkassk in 2007 (power 330 MW). More Russian power generating companies intend to carry out the similar projects.

4.2.3 *Improving the Quality of Coal Fuel*

Production of slurry coal particles in water (coal-water slurry) is one of the important methods of coal-fuel quality increasing. Water-coal fuel (WCF) was invented in the 50s simultaneously with the coal hydro-transport technology. Further improvement of coal-water slurry technology resulted in invention of the new type of fuel known as "artificial compounded liquid fuel" (ACLF).

ACLF has a lot of advantages compared with traditional coal fuel:

- high firing effectiveness,
- less level of toxicity and explosion hazard,
- less waste during transportation,
- high stability features,
- Less emission compared to emissions when burning coal and oil.

One of the most significant goals for coal PE is considered to be its efficiency improvement, since various economic parameters of different branches and environmental conditions totally depend upon it. Even simple adoption of recently developed and used technologies will help to increase the efficiency factor of Russian power plants from 34–40 to 37–47% (lower limit is relevant to average value, upper limit to advanced patterns). Apart from spotlighted tendencies named in the report and concerning coal PE and its efficiency improvement, it is of ultimate importance to get a benefit from positive results of cogeneration usage and optimization of the main power plants machine operation based on minimum fuel cost criterion.

In this way, modern level of coal technologies development in the PE allows significant increasing in coal share in the energy balance and at the same time, solving economic and environmental problems. This conclusion is confirmed by the large difference in the efficiency of coal-fired power plants in different countries, Fig. 4.3. It is promotion to the level reached by advanced countries can provide enormous positive effects in the world as a whole [7].

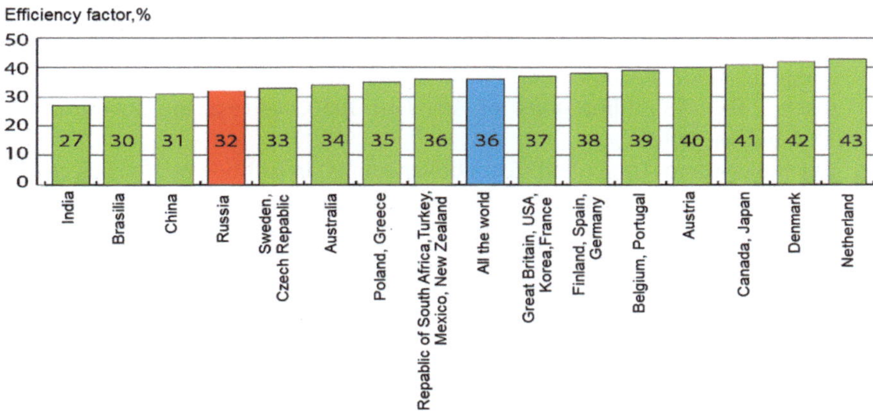

Fig. 4.3 Efficiency factor of coal-fired power plants

4.3   Cogeneration

The traditional way to increase the efficiency of thermal power plants and boiler plants and solving the problem of heating in Soviet and Russian electrical power engineering was and remains the simultaneous production of electricity and heat. In the West, where the problem of heating is not so topical, combined method of production of both types of energy, called "cogeneration" began to evolve relatively recently, but according to forecasts by the association Cogen Europe, the share of cogeneration systems in power generation on average in the European countries in the coming years will reach 20%. Trends for the previous 10 years are shown in Fig. 4.4.

There are several ways to implement cogeneration idea:

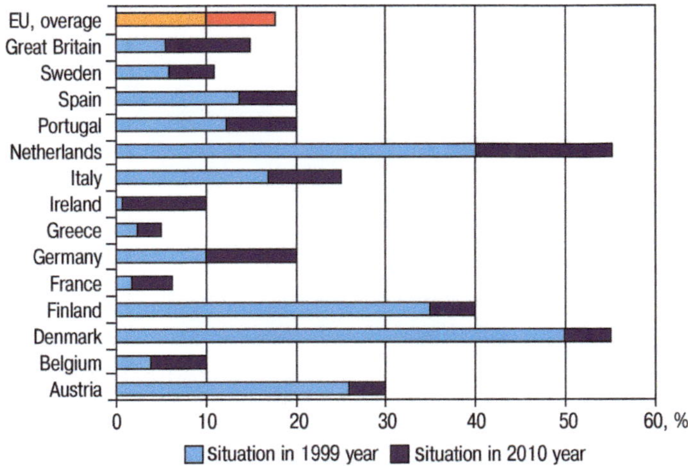

Fig. 4.4 Perspective of cogeneration in the EU

1. At condensation power plants (CPP) intended to produce only electricity, utilize the heat generated in the capacitors and dissipated in the environment. After complete modernization of CPP, the annual fuel saving in the electricity will reach 23% (58 million tons of standard fuel), and saving investments— 15.5 billion dollars.

2. The boiler-rooms (first of all steam boiler-rooms) in addition produce electricity, for example, after substitution of the pressure reducer steam for pressure turbines. The consumption of fuel for generation of 1 kWh of electricity is twice lower than in conventional power stations (140–150 g against 335–345 g of standard fuel). Electrification of the boiler-rooms has a capital intensity around 400 USD/kW, while new capacity in power—not less than 1000–2000 USD/kW.

3. In boiler-rooms with boilers (using fossil fuels, energy of the Earth's bowels, or other sources of hot water) to generate electricity, as used in these hydro-steam turbines operating on hot water which boils in the expansion nozzle and is ejected as a two-phase mixture of these nozzles arranged on Segner wheel. The two-phase mixture rotates it by a jet effect.

4. Boiler-rooms equipped with a superstructure in the form of a gas turbine plant, causing them converted into mini- CHPP (combined heat and power plant). This increases the efficiency of the boiler and gas turbine.

5. Generation of small power plants (diesel engines, internal and external combustion engines, gas turbines, and gas-piston engines) is equipped with waste heat recovery system for production of technological heat, Fig. 4.5.

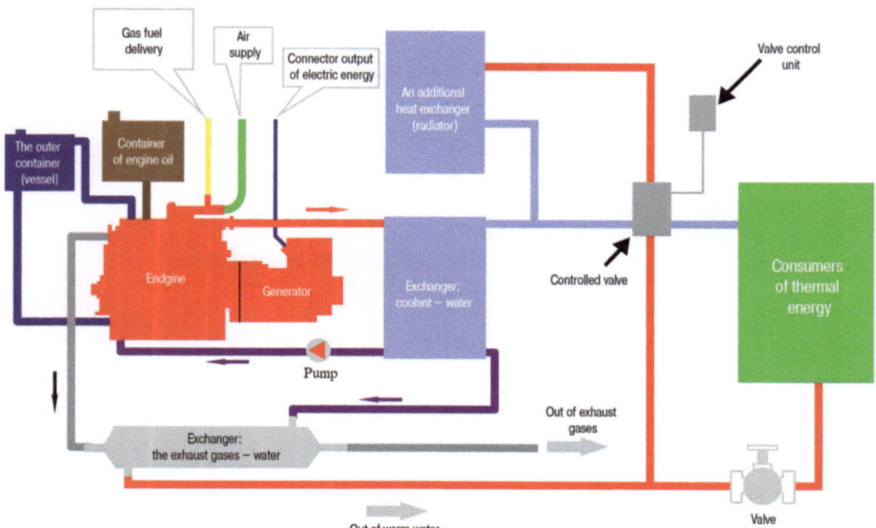

Fig. 4.5 Cogeneration in small power plants

This type of cogeneration is of particular interest to housing and commune sector (HCS), since it reduces construction costs of communications by 1.5–4 times as compared with total district heat and electricity;

6. Powerful gas turbines are set to combined cycle—teamwork gas turbine and steam turbine plants.

A comparison of fuel efficiency for separate and joint production of electricity and thermal energy is shown in Fig. 4.6a–c.

The passing electricity generation by stimulated production of heat (due to the harsh climate in most parts of Russia) is an important area of restructuring of the Russian power industry. In accordance with the ES-2030, due to the electrification of boilers 1.8 trillion kWh of electricity will be produced and the efficiency of its generation will reach 87–88%. The new cogeneration can be regarded as going beyond its powerful energy system into a distributed autonomous power on the

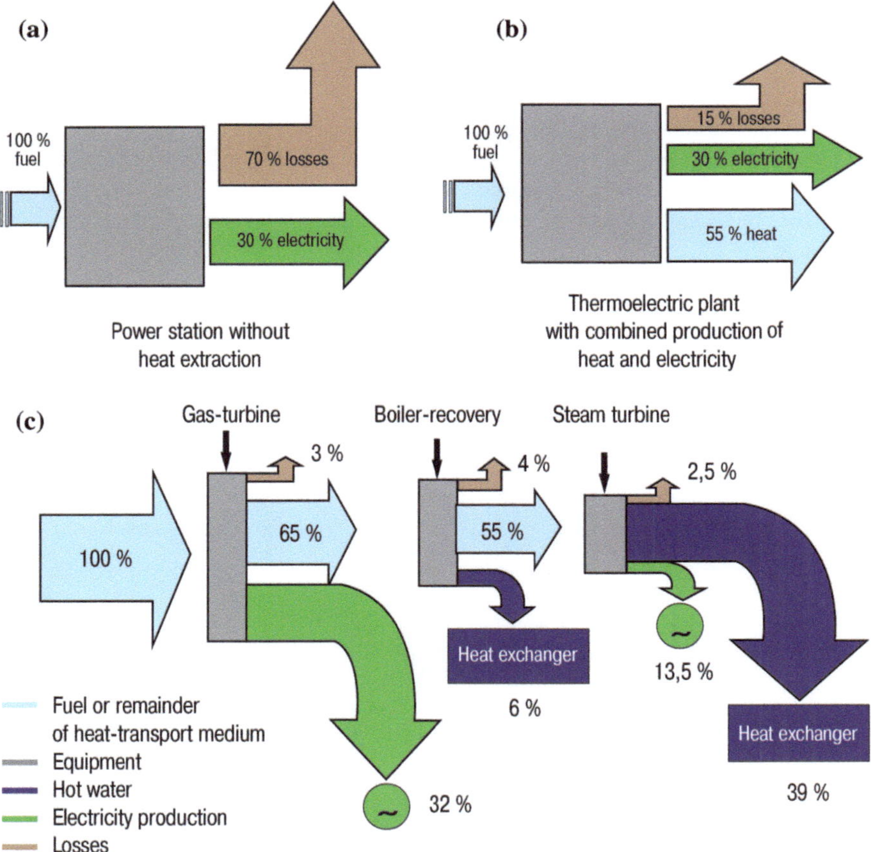

Fig. 4.6 The co-production of the electric and thermal energy

basis of traditional fuels and on renewable energy. The benefits of cogeneration based on the GPI can be shown most clearly in housing and commune sector.

4.4 Small-Scale Power Generation: Current State and Prospects

4.4.1 Distribution Areas of Small-Scale Power Generation

In the world, many millions of people are not adequately supplied with central power and consequently suffer from power and heat energy failures. For example, in Russia about 15 million people, populating 35% of the territory of the country, are not adequately supplied with central power. Most territory is located in the North of Russia, its eastern part. More than 70 cities, 360 urbanized areas, and about 1400 settlements are situated in this area. Its primary sources of electric power and thermal energy supply are more than 6 thousand diesel power plants (DPP) and about 5000 thousand minor boiler plants, consuming more than 2 million tons of coal (Table 4.2).

In whole Russian minor off-system sources of power energy production (power less than 1 MW) nowadays produce only 0.8–1.5% (according to various information sources) of its total volume. The situation with the heat supply differs; however, one can say that in this sphere centralized large-scale heat producers dominate.

The second problem for the regions not supplied with the centralized (system) electricity is the cost of kilowatt-hour. The cost of diesel fuel which is delivered to

Table 4.2 Description of a system of power and heat decentralized consumers in the Northern Russian regions

Power supply	
Number of minor boiler plants	6000
Power production (thousand watts (W) per year)	6.1 (5.6% of power production in the North)
Fuel consumption (million tons per year)	2
Heat supply	
Number of minor boiler plants	5000
Heat energy production (million Gcal per year)	50.2 (31% from heat production in the North)
Fuel consumption (million tons per year)	14
Cost of annual fuel delivery to decentralized consumers	500 billion rubles (20% of budget expenditure of the Northern regions)

settlements by motor off-the-road transport, river barges, and helicopters reaches many hundreds of dollars per ton. That results in high net cost of electric power.

(For the following Russian territories such as Republics of Altai and Tuva, regions of Kamchatka, Magadan, and Chita, Autonomous Areas of Taymyr, Chukchi, Koryak, Nenets, and Yamal-Nenets, and also for some other territories the cost of fuel approaches their six-month budgets.)

For such territories of Russia and others countries, the most attractive way of small-scale power generation development seems to be renewable energy sources [8, 9]. However, according to the energy strategy for Russia and others countries up to 2020–2030 and judging by the real pace of increase in utilization of unconventional renewable energy sources, mineral fuel will be the main power resource for small-scale power generation during several decades (probably till 2040–2050).

In the nearest future, the production methods of power and heat energy out of mineral fuel at power plants of low power will be able to make low (distributed) energy sector competitive in comparison with system energy sector. Their competitive advantages are:

- short construction period (9–18 months),
- quick payback (from 5 years, provided that only power energy is produced, to 2–3 years with full heat utilization in conditions of cogeneration and three-generation (combined cooling, heating, and power),
- Proximity to the consumer and, as a consequence, absence of necessity for large-scale power and heat networks in which dozens percentage of electric power and heat are lost. Today charges on connection of a new house to centralized sources of heat supply are comparable to capital expenses for installation in it by a mini boiler house. The important advantages of small-scale power generation are modularity of generation systems, scalability, and mobility, i.e., capacity delivery to their blocks to meet consumer needs, opportunity for fast connection of new blocks to already working station, and also their dismantle and moving to new objects.

To number of the factors stimulating the development of small-scale power generation also concern:

- adaptation of both consumers and power structures to market uncertainty of power development and prices for power resources that promotes increase in power safety and decrease in investment risks;
- development of new highly effective technologies;
- Growth in the share of high-quality power resources, first of all gas, in energy saving.

There are few experts who consider that installations of small-scale power generation can replace out-of-date thermal power stations and condensation power plants park resources, and that in completely installed gas overpopulated regions of the world the existence of high power installation is justified neither economically,

Table 4.3 Technical characteristics of the most common the gas-fired mini-CHPP

Power (electr.) (kW)	Power (heat) (kW)	Fuel expenditure (kg/h)	Weight (kg)
5	200	80	1500
16	610	250	5000
30	1200	400	8000
120	1500	800	1500
200	2800	1400	28,000

nor technologically. Some of the most common specifications of the gas-fired mini-CHPP are given in Table 4.3.

In developed countries, to generate heat and electricity by direct combustion of wood, wood waste, straw, and peat mini- CHPP are used. In the steam turbine version, the efficiency is 20–25% and power changes from a few kilowatts (for farmers) to hundreds of kilowatts.

In the last few years, the increasing number of the advanced countries chooses as the main way of development of power transition to the distributed power generation. For example, in Belgium centralization of power now does not exceed 20%, though just recently it reached 40%. Experts consider that for Russia a decrease in centralization by 10–15% in the nearest 10–15 years would be optimum. Alongside with other benefits, this will increase efficiency of gas utilization.

The system of gas utilization developed in Russia is extremely irrational: the most part of gas is spent on thermal energy production (64%), the significant share of the electric power is produced on condensation power plants with low efficiency factor; the huge thermal power stations, which are not completely loaded with heat production, are forced to work in a condensation mode that essentially reduces their efficiency factor.

In the market for small-scale power generation there are specialized companies manufacturing highly effective equipment and having partners the investment and leasing companies. Their customers are enterprises and municipal unions, oilmen and gasmen, builders, owners of summer residences, cottages, farms; rescuers and firemen, etc.

Systems of independent power supply (IPS) are applied in three situations:

(a) In the absence of the centralized power supply, consumers have only one choice—to what specific kind of IPS gives their preference (traditional fuel, nonrenewable power-generation resources, or two-three diverse energy assembly combinations);
(b) With poor quality of electric supply (frequent switching-off, fluctuations and voltage slumps);
(c) In the presence of competitive advantages of IPS.

Thus market laws inure: the consumer makes the choice of a power supply source on the basis of an estimation of his/hers consumer qualities.

Emergency power supply system construction for a private sector is economically proved both in areas of mass building and outside of cities. In the first case, the additional effect is achieved due to economy of means on capital construction old boiler-houses and boiler rooms can be converted into thermal mini power plants. In the second case—lining of power networks and connections to them are not necessary. Their specific cost reduced to 1 kWh of the established capacity compared with expenses for installation of a cogeneration unit differs sensationally expenses on cogeneration unit are compensated within 3–5 years due to a difference between the tariff and the cost price, and expenses on connection to the centralized source are irrecoverably lost by transfer of newly constructed substations on the financial statement of the power companies. The new order of payment collection for connection to the energy source was introduced in 2005; it makes power supply from emergency power supply system even more attractive. They can be used as basic, as well as emergency, reserve, and meeting the needs for the electric power at the time of peak loading, they also can be used for associated gas recycling and production of electric energy alongside with thermal energy.

The special urgency in conditions of restriction of the maximal power consumption and action of double-rate tariff is getting the use of emergency power supply system to meet the needs for the electric power at the time of peak loading. It is more economical to users of the given units due to power consumption reduction at the time of peak loading from the centralized source at the price that essentially exceeds the daily average.

In the last few years, the increasing popularity has been getting mini-boiler-houses installed in apartments and detached houses; they allow to lower the most unprofitable clause in services of housing and communal services-cost of heat. By power efficiency such heating surpasses almost twice regional boiler-houses and by a factor of 1.2–1.3 independent modular boiler-houses. Also mini-boiler-houses work in a cogeneration mode (the so-called house assemblies), and only gas is supplied to them; electric power and heat are produced inside of most assemblies.

4.4.2 Small-Scale Distributed Power Generation Functioning on Organic Fuel

Development and mastering of electric and thermal energy production methods using low power energy assemblies working on liquid and gaseous fuel are at the center of attention nowadays. Plants generating small power usually include the following:

- Gas-turbine installations (GTI),
- Gas-piston installations (GPI),
- Diesel,
- Gas and diesel fuel,
- External combustion engines—Stirling engines (SE),
- Expander—generating installations (EGI).

4.4.2.1   Gas-Turbine Power Installations

Achievements in the field of creation of turbines for vehicles in aviation, automobile, and ship-building branches have a great influence on development of gas-turbine methods of electric power productions within the limits of independent power. Penetration into power production of highly effective (efficiency factor up to 55–60%) gas-turbine and steam and gas installations (GTI and SGI) of a wide capacity range, including small—from units to one or two tens of megawatts begun in the 1980s. On their basis small GTI heat and power plants for the combined electric power and heat production (mini-CHPP) were constructed. Distribution of the gas-turbine technology in Russia was also promoted by emergence on the market of the five world "competence centers" in this sphere—companies such as "Capstone", "Honeywell Power Systems", "Kawasaki", "Bowling Power System", and "Eliot Energy Systems".

Russia can also take a worthy place in the world gas turbine technical equipment market owing to the last 2–3 decades of successful development of science and production association "Saturn" and more than ten other enterprises, basically, enterprises of the former USSR "Aviaprom". "Saturn" sells a complex of industrial programs, including those in the sphere of power mechanical engineering, focused on needs of electrical power engineering in Russia, gas and oil companies, and municipal unions.

The favorable conjuncture for small-scale power generation and rapid technical and technological progress in this area make GTPI-heat and power plants more and more attractive. Terms of commissioning of such power plants, owing to high GTI factory readiness, make about 1 year. Costs of produced energy at GTI-heat and power plants in many regions also appear competitive with large power stations. Their competitiveness essentially increases if to combine gas turbines with steam; this provides the efficiency factor close to 70%. GTI in the cogeneration mode has effective operating ratio of fuel of 90%. Thus, from 1 kWh of electric capacity, 1.3 kWh of thermal is produced.

The main advantages of GTI in comparison with steam-turbine installations (STI) are:

1. Low specific weight, compactness and simplicity of transportation and installation; modern gas turbines (especially with capacity not exceeding 16 MW) are available in the form of one or more blocks of full factory readiness, requiring a minimum amount of installation works, and make low demands to construction works and infrastructure.
2. Small amounts of harmful emissions into the environment.
3. Possibility of rapid replacement of the gas turbine units.
4. Relatively low capital investment and low (for power plants) payback period.
5. High-speed maneuverability and load set; even for large GTI time to tens of minutes is required to reach full power, in contrast to the STPI whose start takes dozens of hours.

6. Most of the GTI permit of overload i.e. increases in power above the nominal. This is achieved by raising the temperature of the working fluid. The duration of such conditions should not exceed a few hundred hours in order to avoid significant reduction in installation resource.
7. High economic efficiency of the GTI is achieved through cogeneration (creation of the GTI-CHPP), when they produce not only electricity but also heat (heating, domestic hot water, heat supply for production needs). GTI in cogeneration mode have a fuel efficiency of 90% and the same time produces 1.3–2.5 kW of heat per 1 kW of electrical power. In cogeneration mode the electricity cost is reduced by 20–45%, which makes the GTI-CHPP in many regions competitive with the large power plants in terms of cost of energy produced [10]. The payback period is 3–4 years.

It is remarkable that in many countries, using in energy sector significant volumes of gas, other technologies are forbidden. The increase in the share GTPI-heat and power plants in the EU countries is illustrated by Table 4.4.

In Russia small GTI heat and power plants now appear effective not only in the remote but also in developed regions, Fig. 4.7.

Optimistic views concerning the future of small GTI heat and power plants inspire high interest to them from oilmen and gasmen, experiencing greater needs for heat and electro supply and having sufficient financial opportunities. Gas-turbine technology allows utilizing associated gas, provides the lower cost of produced energy, and cuts down expenses on environmental protection.

Expansion of gasification sphere on both middle and small cities and settlements provides their active involving in generation of capacity in many regions of the country. In the long term the construction of small GTI heat and power plants instead of the uneconomical and out-of-date boiler-houses in cities and towns will make 25–35 GW to 2020 and 35–50 GW to 2050, i.e., up to 10–15% of the total established capacity.

GTI has a number of disadvantages, narrowing the scope of their competitive advantage:

- Relatively high requirements for the quality of the gas fuel, associated with the need to prevent the high-temperature corrosion of turbine blades (restrictions are usually imposed on the total content of sulfur and alkali metals);
- Need for pre-compression of gas fuel significantly increases the cost of energy produced (especially for small gas turbines) and in some cases prevents their implementation in the energy sector. For modern GTPI with high degrees of air

Table 4.4 Predicted dynamics of the implementation of GTI-CHPP (mostly small capacity) in the EU

Years	Capacity GW	The share in total installed capacity (%)
2000	74	12
2010	91–135	13–18
2020	124–195	15–22

Fig. 4.7 Gas turbine power plant using associated gas (capacity of 24 MW, owned by "Gazpromneft-Vostok", Russia)

compression, the required fuel gas pressure may exceed 25–30 kg/cm^2 (fuel gas pressure must exceed the air pressure by not less than 10.5 kg/cm^2);

- Sharp drop in the efficiency at lower loads, especially inherent in HD-GTI (heavy-duty GTI); AD-GTI (aero derivative GTI) have more advanced control mechanism of the compressor blading that partly smoothes this deficiency; moreover, the free variable speed shafts also support multiple levels of efficiency. However, the efficiency of gas turbine engine loads below the 60–50% is problematic;
- Service life of GTI is considerably smaller than that of other power plants and is generally in the range of 45–125 thousand hours (the lower limit applies to forced AD-GTI). There is significant progress in increasing of the service life of GTI.

4.4.2.2 Piston Installations

Historically, during many decades and up to now in the majority of small power installations are used piston engines, working on liquid fuel (mainly diesel engines). As the share of gas in the general volume of power raw material began to increase, the share of gas-piston installations (GPI) also began to increase.

Currently, the GPI successfully compete in the low-power engineering with diesel piston engine and with GTI due to their existing benefits:

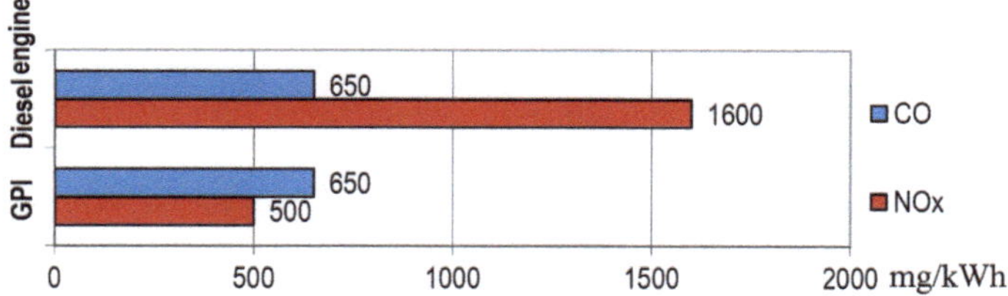

Fig. 4.8 Level of harmful emissions during operation of a diesel engine and the GPI

- the slower rate of increase of gas fuel cost compared with liquid;
- significantly less harmful emissions of gas-fueled engine compared to liquid-fueled engine, Fig. 4.8;
- rapid development of the gas network;
- low cost of the installed capacity of the unit; CHPP capacity of 10 MW on the basis of the GPI requires investment of about $7.5 million, and based on GPI— about $ 9.5 million;
- The best indicator of specific weight (power to weight ratio). The reason is that the equipment provides the gas supply to the engine much easier than the complex aggregates associated with the storage and supply of fuel oil;
- higher efficiency factor less dependent on the load;
- Specific charge of GPI fuel with the capacity of one cylinder more than 15 kW practically does not depend on capacity. This means that to ensure of flexibility of the energy production, it is expedient to build at one enterprise several mini-TPP on the base of GPI.
- Security—the lack of high temperatures, pressures, moments of inertia;
- long service life of GPI—up to 300,000 h or 37 years in operation for 8000 h per year, making mini-CHPP reliable uninterruptible power supply;
- GPI can be started and stopped any number of times, that does not affect the overall life of the engine, while 100 GTI starts reduce resource for 500 h;
- time before the load after the start of the GPI is 2–3 min, while of the GTI—15 to 17 min;
- lesser dependence of the efficiency factor of the GPI on the ambient temperature;
- mobility;
- a wide range of operating modes—from 15–20 to 110% (in peak mode) rated capacity in proportional fuel consumption;
- short payback period—3 to 5 years;
- autonomy of mini-CHPP based on them, producing electricity and heat at the point of consumption and giving a guarantee against interruptions, outages, or additional energy losses;
- the possibility to solve the issue of uneven daily electricity consumption, unsolvable for large generating plants relatively easily;
- "omnivorousness"—the GPI can run on natural gas, propane, butane, associated petroleum gas, coke oven gas, biogas, etc.

The global market for stand-alone power plants is characterized by a wide variety of required power at different weight, size, and resource base indicators of generators. There is a shift of demand from the emergency power in the direction of power for continuous operation.

It is economically preferable to construct mini-CHPP with the combined energy production on the basis of GPI with capacity up to 3 MW and on the basis of GTI—with capacity 30–50 MW. Most widespread mini-TPP located in developed countries work on the basis of GPI with individual capacity from 30 kW to 12 MW. GPI manufacturers and experts in the field of the power equipment consider the following three especially perspective spheres of their application: associated gas usage, meeting own boiler-house needs in the energy sector, production of electric and thermal energy for housing and communal services needs. The tendency of displacement of GPI use from emergency unit's aside power stations for permanent operation also should be mentioned.

Among the disadvantages of piston machines it is necessary to note the following.

1. Limited capacity is 5 MW for a single machine. However, this is not critical, as several units working in parallel can be installed, if necessary. There are examples of installation of up to 40 units in a local system. In addition, the Finnish company Wartsila produces powerful GPI with a unit of electrical capacity of over 16 MW.
2. Increased need for lubricating oil in comparison with gas and steam turbines. Lubrication problem for the GPI (as for diesel engines) is converted not only into technical, but also in economic problem. For production of 50–200 kW power oil consumption can reach 2.5–3 g/kWh, which is a significant proportion of total operating costs, particularly if the engine runs on the cheap gas fuel.

4.4.2.3 Stirling Engine

Stirling engine (SE) is the most well-known representative of the family of external combustion engines—an external combustion engine that uses air or an inert gas as the working fluid operating on a highly efficient thermodynamic cycle, Fig. 4.9. The SE produced by such companies as "Philips", "STM Inc.", "Daimler Benz", and "United Stirling" with the capacity from 5 kW to 1.2 MW the have efficiency factor more than 42%, the working resource more than 40 thousand hours, and the specific weight from 1.2 to 3.8 kg/kW. Their characteristics and consumer qualities surpass internal combustion engines (ICE) and GTI.

The SE-based technology can also suggest advantages over turbine-based systems in low temperature differential applications concerning to low-cost materials, fabrication, and maintenance. There are the research programs looking at the potential viability of SE-based power generation systems utilizing low grade heat sources, where temperature differentials down to around 30 K may be considered.

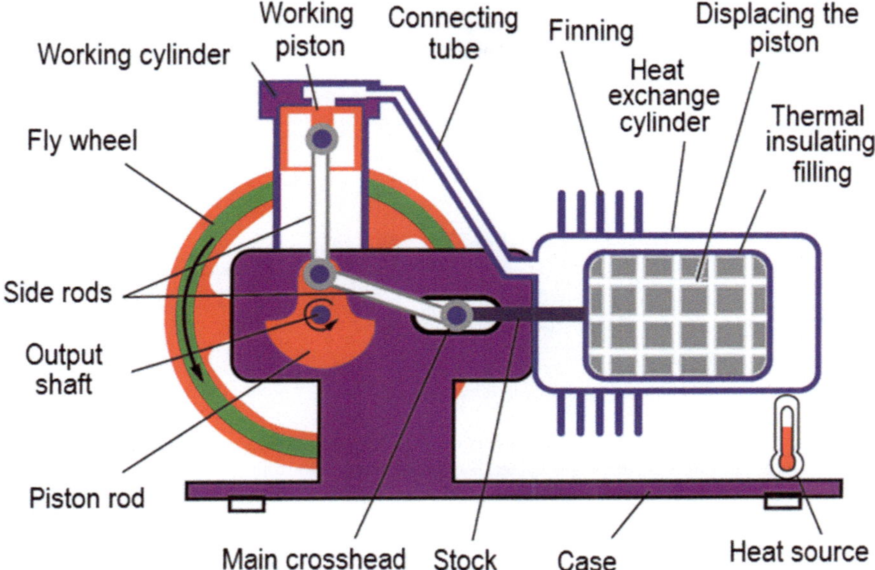

Fig. 4.9 One of the possible schemes of the Stirling engine unit

The SE has the following characteristics: low noise level, almost two times smaller concentrations of harmful substances in exhaust gases (in comparison with ICE), greater resource, and good characteristics of the twisting moment. It is very important that SE can work both on traditional liquid motor fuel and on gas of any origin, and also on a gas condensate. This means that spheres and schemes of its use are about the same, like other independent energy units.

Despite the above listed advantages, SE has not found a large-scale application yet, owing to several problems of technical realization of this tempting and, for a long time (since 1816), known idea. The problems are as follows: great difficulties in building of some units (condensation, regulation capacity systems, etc.), high requirements to the production methods, necessity of application of heat-resistant alloys and metals, and new methods of their welding and soldering. First of all these problems are caused by application of such working substances as helium and hydrogen.

4.4.2.4 Turbo-expander Generators

Turbo-expander installations are a variety of gas turbines; they are designed for recycling of heated gases or gas energy under elevated pressure for electricity production. They are mainly used as ***turbo-expander generators (TEG)*** with decreasing gas pressure in the main and distribution pipelines to the desired user.

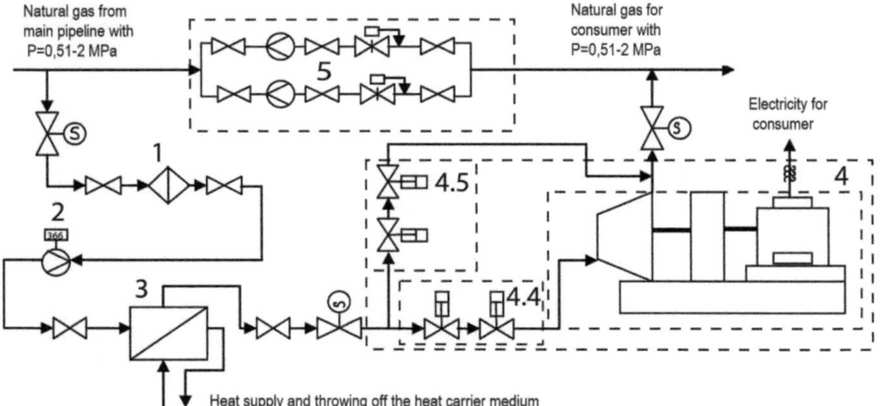

Fig. 4.10 Technological scheme of connection of a gas distributing point GDP to the expander unit: *1* filter; *2* gas flow meter; *3* gas heater; *4* expander-generator unit: *4.1* expander, *4.2* generator, *4.3* gearbox, *4.4* flow metering valve, *4.5* pressure regulator block for the bypass line; *5* gas reducing points

They are promising for the production of electricity in TPP, and a number of other industries, where the generated via TEG electrical or thermal energy is a byproduct: in metallurgy (for reducing blast-furnace gas pressure), chemical plants, liquefaction technology, separation of gases, air conditioning, and others.

Leading European countries and the USA have been using this almost free source of energy for several decades. In Russia, its use began only in the last 10–15 years. And this despite the fact that the idea of using high-pressure gas into gas pipelines for generating electrical energy was first proposed by the Russian academician M.D. Millionshchikov in 1947. The technology is based on the fact that parallel to gas reducing points of transmission and distribution pipelines, turbo expanders are installed. The latter, acting as gas distributing points (GDP) and gas distributing stations (GDS), generate electric power, Fig. 4.10.

The appearance of the expander-generating units is shown in Fig. 4.11. Relatively small expander-generator sets, installed on the GDP and GDS distribution network, should be used to produce not only electricity, but also cold. This makes possible the construction at GDS industrial refrigerators, the capacity of which will be determined by the value of a stable gas flow through the expander. Most TEG with a unit capacity of 1.5–6 MW can be considered acceptable.

The preliminary study of the project of energy-technological expander plant based on the GDS with a stable daily gas consumption of 60 thousand m^3 (Fig. 4.12) shows that its cooling capacity is sufficient for a typical industrial refrigerator with a capacity of 270 tons. In this specific case production of electricity in the installation of 0.025 kWh/Nm^3 and electric generator power of 62.5 kW are sufficient to meet their own needs refrigerator (automation, pumps, lighting, etc.).

Fig. 4.11 Photo of the expander-generating units

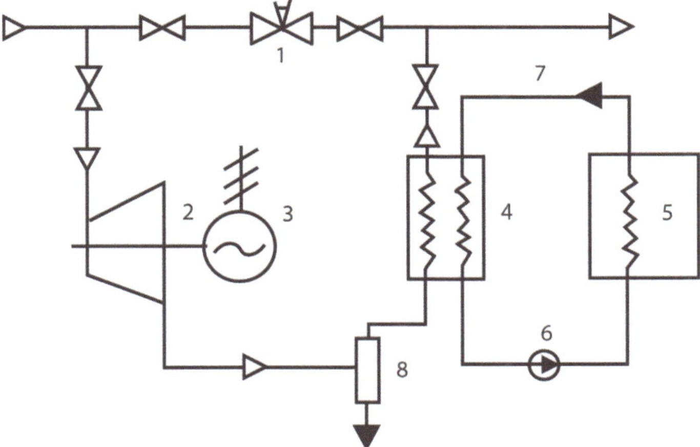

Fig. 4.12 Concept of energy technological expander plant: *1* reducing valve GDS; *2* screw-shaped expander; *3* generator; *4* heat exchanger; *5* freezer; *6* circulating pump; *7* refrigerant circuit; *8* separator

Payback period of projects is from 3 to 5 years. Market energy technological installations using excess gas pressure drop across the relatively small GDS and large GDP to generate electricity and cooling of industrial and agricultural refrigerators are also great. The main advantages of TEG are:

- lower specific capital costs;
- lower energy consumption for electricity generation associated with heating of the gas before the expander heat is released during the combustion of fuel.

The process of gas expansion in the expander can be arranged in such a way that almost all the gas supplied to the heat is converted into electricity. If you use gas for heating or high potential of renewable energy sources (RES), heat pumps can produce electricity without any fuel combustion in TEG:

- High environmental performance (lower volume of emissions into the environment in comparison with other power plants). This is due to a lower specific fuel consumption for electricity generation (in the limit-up to zero);
- The use of TEG in the TPP and industry, for which electricity is not the main product, can improve their technical and economic performance by reducing energy costs.

On the TPP, for which electricity is the main product produced, the use of TEG improves the technical and economic indicators of the main production (reduction of specific fuel consumption for electricity generation). In the event that, for whatever reasons, there is no possibility of organizing a flow of electricity from one grid to another, a necessary condition when the TEG is connected to heat circuit CPP is to keep the power generation station as a whole at the same level as it was before the inclusion of TEG. That is possible due to reduction of electric power generated by the steam turbine units and the amount of electric power generated by TEG.

TEG can be switched on into the thermal circuit of the CPP operating with a deficit of electric capacity. The nominal flow rate of steam to the turbine in this case is ensured, Fig. 4.13.

4.5 Nuclear Power Engineering

4.5.1 Current State and Prospects for the Development of Nuclear Power Plants with Uranium Fuel Cycle

About 438 nuclear power units (about 194 nuclear power plants) operated in 32 countries all over the world in the middle of 2016. They generated 400 GW of electric power (the largest nuclear power station is Kashiwazaki Kariva (Japan) with 7 nuclear power units generating P_{tot} = 8200 MW). The largest number of nuclear reactors is situated in the USA (104), France (58), Japan (51), Russia (34) and the Republic of Korea (23). 61 nuclear power units with total installed capacity of 70.6 GW (el.) are under construction. About 300 research and experimental nuclear reactors operate in 56 countries. More than 200 nuclear reactors are installed in ships and submarines.

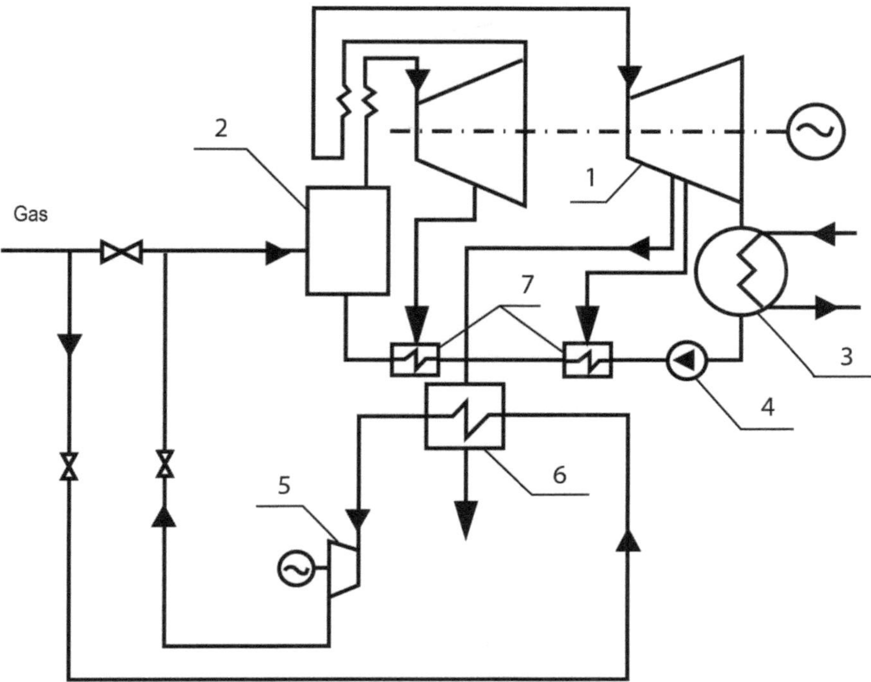

Fig. 4.13 Scheme of inclusion of TEG in thermal power scheme: *1* turbine; *2* boiler; *3* condenser; *4* pump; *5* expander; *6* heat exchanger for gas heating; *7* regenerative heaters

The largest shares of atomic power engineering in the total power engineering have France (76.1%), Belgium (55.5%), Ukraine (46.8%), Sweden (46.2%), Armenia (45.5%), Bulgaria (45.4%), Slovakia (44%), Switzerland (40.6%), Slovenia (40%), and Hungary (40%). APS generate approximately 17% of the global electric power, Table 4.5.

The total generated electric power increased, because old and small NPP had been taken out of service and new and large reactor units were put into service, Fig. 4.14.

Strengthening the APE positions allows nuclear engineers to acquire more and more abilities to pay attention not only to the first-level priority problems—safe operation of NPS and burial of radioactive wastes and spent nuclear fuel—but also to innovations in a wide spectrum of directions in the development of technology including:

(a) Modernization of the design of high-power reactors and power generating units and technology of their construction,
(b) Development of nuclear central heating,
(c) Development of low-power nuclear stations (LP NS), including floating ones intended for power supply of remote territories,

Table 4.5 Main characteristics of world atomic power engineering

Country	Capacity (MWe)	Share (%)	Number of plants	Number of units
Japan	44,215	29.21	17	50
South Korea	20,618	35.48	6	23
Republic of South Africa	1800	5.18	1	2
Sweden	9298	38.13	3	10
Switzerland	3263	38.01	4	5
Czech Republic	3678	33.27	2	6
France	63,130	74.12	19	58
Finland	2716	28.43	2	4
Ukraine	13,107	48.11	4	15
Taiwan	4927	19.30	3	6
The USA	101,240	19.59	66	104
Slovenia Croatia	688	37.30 8.0	1	1
Slovakia	1816	51.80	2	4
Romania	1300	19.48	1	2
Russia	23,643	17.81	10	33
Pakistan	725	2.60	2	3
Netherlands	482	3.38	1	1
Mexico	1300	3.59	1	2
China	11,688	1.92	7	16
Canada	12,624	15.07	5	18
Spain	7567	20.09	6	8
Iran	915		1	1
India	4391	2.85	6	20
Germany	12,068	28.38	8	9
Hungary	1889	42.10	1	4
Great Britain	9703	15.32	8	17
Brazil	1884	3.06	1	2
Bulgaria	1906	33.13	1	2
Belgium	5927	51.16	2	7
Armenia	375	39.42	1	1
Argentina	935	5.91	2	2
All over the world	369,818		194	436

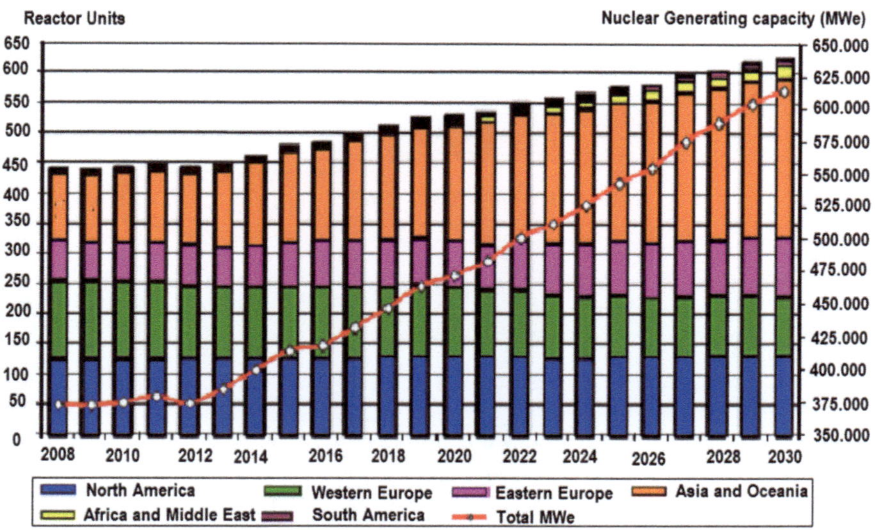

Fig. 4.14 Forecast of development of nuclear energy (dates and regions)

(d) Mastering of the closed fuel cycle based on fast-neutron (FN) reactors,
(e) Hydrogen production by the water thermolysis in high-temperature atomic reactors, etc.

Over the past few years, a severe problem of atomic power engineering has become the protection of atomic objects against acts of terrorism (in particular, after acts of terrorism in the USA on September 11, 2001 and in Russia in October 2002, origin of the Islamic State of Iraq and the Levant, 2012).

The nuclear power stations with reactors on thermal neutrons operate in the mode of open nuclear fuel cycle including three stages:

The 1st stage begins from extraction of uranium ore and ends with the delivery of fuel assemblies to nuclear power station site,

The 2nd stage begins from fuel utilization in the reactor to generate electric power and ends with temporal storage of SNF on the NPP site,

The 3rd stage begins from SNF dispatch to a special storehouse and ends with nuclear waste burial, Fig. 4.15.

The burial and utilization of radioactive wastes from nuclear objects (NPP, nuclear ships, submarines, etc.) are also severe problems. Members of Greenpeace adduce arguments confirming a threat to the biosphere from such wastes. Atomic physicists adduce opposite arguments and simultaneously improve systems of utilization and burial of radioactive waste.

The modern concept of NPP safety is based on three principles: control, multi-level protection, and technical engineering means of safety. More than 20 projects of NPP of new generation that differ radically not only in the power and reactor type but also in the technological, schematic, and design solutions are being

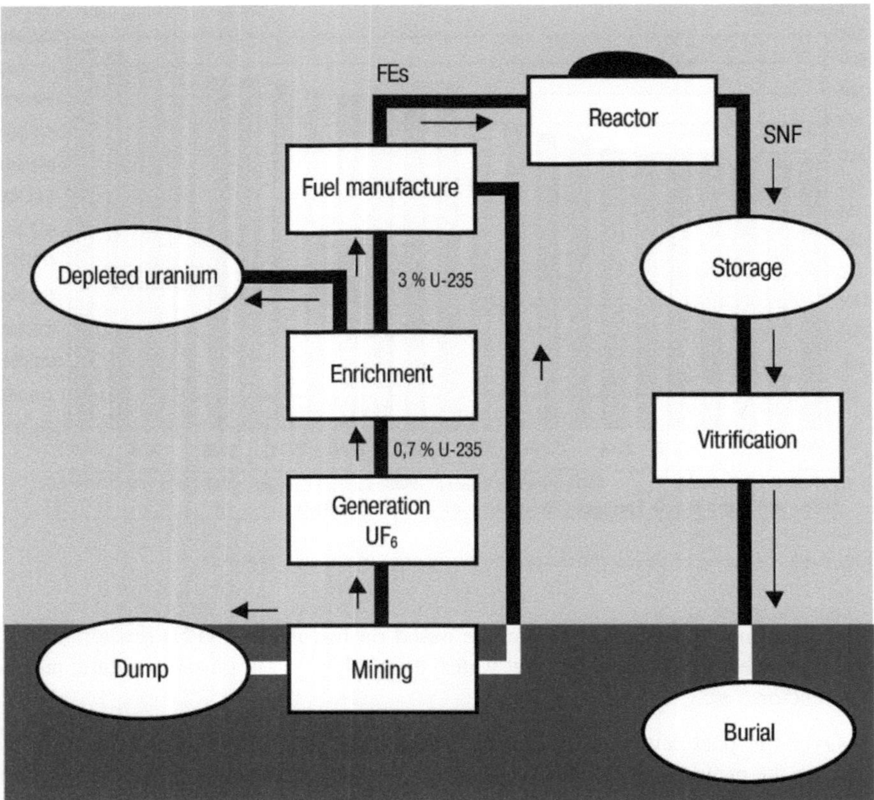

Fig. 4.15 Open nuclear fuel cycle

developed by leading energy corporations and companies of industrially developed countries.

Based on terms of commercial realization and degree of self-safety, NPP are conditionally subdivided into three generations. The projects of NPP of a new generation are based on technology and design that have already been mastered and checked under operating conditions. They are equipped with active and passive safety systems that allow the probability of serious accidents to be decreased and the investment price of electric power generation to be reduced by 20%.

Antinuclear motion in many countries has forced governments of some of them to refuse completely from atomic power engineering (for example, in Germany) and even to close operating NPP. Vice versa, other countries (for example, France) continue the development of atomic power engineering and increase the share of electric power generated by NPP.

As a whole there is a positive revaluation of the role and place of atomic power engineering in provision of the energy safety with improvement of reactor design and technology of the entire nuclear fuel cycle. There are several reasons for this:

Cost of generated electric power; nuclear power stations are quite competitive with coal and gas power stations;

Cost of electric power generated by NPP is less dependent on conjuncture of the price for fuel due to the fact that uranium is quite accessible and the share of the fuel cost in the cost of the electric power generated by atomic power stations is smaller than that in the cost of the electric power generated by coal and gas TPP;

New more reliable reactors change favorably the public opinion about nuclear power stations;

Considerable stocks of nuclear fuel and absence of problems of its delivery to the NPS make APE an important factor of ensuring regional, national, and global energy safety (one fuel tablet of uranium dioxide provides the amount of energy equal to that from combustion of 882 kg wood, 550 kg coal, or 500 kg oil);

Absence of emissions of "greenhouse gases" and solid particles makes the nuclear power station out of competition with coal and gas power stations by ecological compatibility. (The coal power station generating 1000 MW per year consumes 3 million ton coal or 75,000 railroad cars and 5×10^9 m^3 oxygen; in this case, about 0.7 million ton of solid wastes are produced.)

Strengthening of the NPE positions allows nuclear engineers to acquire more and more abilities to pay attention not only to the first-level priority problems—safe operation of atomic power stations and burial of radioactive wastes and spent nuclear fuel—but also to innovations in a wide spectrum of directions of the development of technology:

(a) Modernization of the design of high-power reactors and power generating units and technology of their construction,
(b) Development of nuclear central heating,
(c) Development of low-power nuclear power plants (LP NPP), including floating ones intended for power supply of remote territories,
(d) Mastering of the closed fuel cycle based on FNR,
(e) Hydrogen production by water thermolysis in high-temperature nuclear reactors, etc.

Nowadays scientists and engineers are intensively working at one more nuclear energy source and increase of the efficiency of nuclear fuel burning. In Russia occupying the leading positions in the field of nuclear technologies, a number of projects on modernization of reactors on thermal neutrons have been developed, are realized, or have already been realized. Among them are

- Completion of works aimed at increasing thermal capacity of operating WWER-1000 and designing of new nuclear power units with thermal capacities of 3200 and 3300 MW.
- The 18-month fuel cycle has been introduced in WWER-1000.

- Works are underway on the substantiation and introduction of maneuverable regimes of operation.
- Positive results of experiments on utilization of regenerated uranium fuel in WWER have been obtained.
- Preliminary investigations of the possibility of utilization of REMIX fuel (mixture of uranium and plutonium) and MOX fuel (UO_2 + PuO_2) are carried out. (Fast neutron reactors can be found in Sect. 4.5).

4.5.2 Intermediate and Low-Power Nuclear Stations Including Floating Ones

Nuclear power stations with electric capacity not exceeding 300 MW and thermal power stations with electric capacity not exceeding 500 MW belong to this class of stations (LP NPP). A lot of effort has been done in nuclear major countries to develop small reactor (with capacity 2.0–35 MW) for enhancement of nuclear peaceful use in district heating and electric power generation in many regions of the world that have no central power supply (in Russia among them are northern and eastern regions of the country, see Sect. 4.4). They can be used for seawater desalination, hydrogen generation and so on.

A wide number of capacities, long period of autonomous operation (for 10–50 years they do not need additional fuel loading), and high degree of pre-fabrication make them especially attractive for these regions. Heat load distribution and its local character and short range of economically efficient heat transfer predetermine dispersed arrangement of thermal energy sources and their maximum capacity commensurable with the value of local loading.

The LP NPP can operate as integral parts of large electric systems. In this case, among their advantages are (a) shorter terms of commissioning and shorter payback period that make them more attractive to investors in comparison with large nuclear power units, (b) smaller financial, radiation, and technogenic risks, (c) large-scale economy of natural gas in the spheres of central heating and heat supply of large cities, (d) preset level of safety of their operation needs much smaller engineering safety systems. This is provided by means of excluding their occurrence as those due to rational application of laws of nature, including negative feedbacks in reactor physics, rather than by means of application of fusers or localizing systems.

South Korea has been developing an integral type nuclear cogeneration reactor. It is a modular pressurized water reactor and it is expected to be used for dual-purpose applications of seawater desalination and small-scale power generation. Since this reactor will be located relatively near the residential area, it should have highly enhanced safety characteristics compared with current nuclear power plants. The electric power capacities of the developed reactor are 2, 5, 10, and 20 MW.

In the last few years, the use of reactor units designed by the technology of shipboard modular reactors to construct nuclear thermal power stations and floating LP NS has been studied that together with small nuclear power units based on traditional fuel and not-traditional renewable energy sources (NRES) would improve living conditions and economic activities in regions of the world far removed from the central electric power and heat supply.

In the field of low-power nuclear reactor design, Russia occupies the leading position. Already in the early 1990s there were more than 40 projects of such reactors of various technology readiness levels. A number of Russian projects have already been supported by the international community including International Atomic Energy Agency (IAEA).

Nowadays more than 50 concepts and projects of low and intermediate power reactors (300–700 MW) of various types, including water cooled, liquid metal cooled, and gas cooled reactors are being constructed in 15 developed and developing countries all over the world. All projects provide increased safety measures of LP NPP operation, and some of them also envisage long-term continuous operation without refueling, thereby improving the protection from the spread of fissile materials allowing one to undertake adequate measures and to guarantee non-proliferation in the scenario of global large-scale APE development. These projects combine the advantages of many well-developed technologies—fuel, cooler, and power converter—that allow reliable and safe operation in any regions, including regions with extreme geological and climatic conditions.

The innovative project "Sources of Electric and Thermal Energy Based on Technologies of Nuclear Shipbuilding" has been developed in Russia. Within the framework of its implementation, the construction of the first floating nuclear power station (FNPP) "Academician Lomonosov" was started in 2007. Two reactors having each 35 MW (electric) and 148 MW (thermal) capacities will be mounted on a transportable barge. This floating platform is 140 m long, 30 m wide, and 15 m high with displacement of 24 thousand tons will sail to the shore to be connected with mainland and then will generate energy, Fig. 4.16. An FNPP will autonomously operate for 4–5 years; its total service time will be about 40 years. Its permanent staff is 78 people working in shifts.

This project is planned as a serial one, providing construction of 7–10 interchangeable stations capable to supply energy to 70% of Russian territory uncovered by central power supply (including mainly coastal lines of Arctic and Pacific Oceans). The level of readiness in the middle 2015 was estimated to be 70%. It is planned to be finished in 2019. The second station will be put in operation in 2030.

Such nuclear power plants can be of interest for island states of the Pacific region to produce drinking water the need in which frequently exceeds that in the electric power.

Experience accumulated in the design of the first FNPP and accident with Fukushima-1 NPP demonstrated that the main FNPP disadvantages are:

Fig. 4.16 Floating nuclear power plant with two 35-MW reactors: *1* storehouse for spent fuel and radioactive waste, *2* reactor units, *3* steam turbine units, *4* foundation pit (9 m deep) filled with water, *5* hydraulic engineering facility, *6* heater, *7* devices for electric power distribution and transfer to the consumer, *8* bunker of wet storage of salt, *9* tanks with hot water

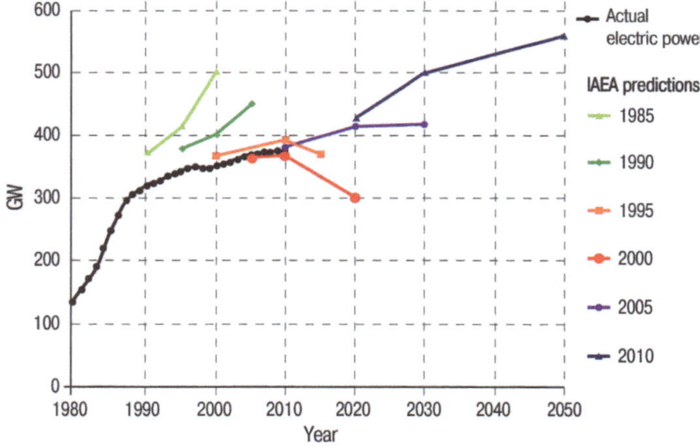

Fig. 4.17 IAEA predicted and actual rates of APE development

High cost of installed capacity unit due to high cost of the floating platform and need of connection with the coast (approximately 0.5 billion USD together with the reactor),

Inadmissibility of application in regions, where tsunami and hurricanes arise.

We must recognize NPE should overcome great difficulties in the process of its development. This is confirmed by the fact that none of the IAEA predictions on the NPE development has come true, Fig. 4.17.

Proponents of the traditional energy fuels assess the current state of NPE as stagnation, underlining that the share of nuclear power stations in the total electric power generation all over the world has been reduced in the last 3–3.5 decades from 17 to 12%.

Nuclear engineers consider this tendency to be temporal and difficulties to be surmountable and distinguish three main stages in the development of atomic technologies in the XXI st century (according to the materials of the National Research Center "Kurchatov Institute"):

1. Nearest stage (10–12 years):
   Evolutionary development of reactors and technologies of the fuel cycle,
   Elaboration and experimental operation of improved and innovative reactor and fuel cycle technologies.
2. The stage of active growth of atomic power engineering (up to the middle of the century):
   Increase of production volume by four-five times,
   Development of innovative reactor and fuel cycle technologies (expanded reproduction of fuel using fast neutron reactors, utilization of useful and burning out of dangerous isotopes, high-temperature reactors, production of hydrogen, fresh water, etc.).
3. The stage of sustainable development of large-scale nuclear power engineering (second half of our century):
   Deployment of innovative nuclear technologies,
   Multi-component nuclear power engineering,
   Nuclear-hydrogen power engineering.

4.5.3   Nuclear Power Plants with Fast Neutron Reactors

Escalating volumes of extraction of natural uranium for the existing NPE technology based on thermal neutron reactors (TNR) (energy of thermal neutrons is 0.025 meV) cannot provide long-term development of large-scale atomic power engineering. With application of only TNR, the resources of the atomic power engineering exceed not so strongly those of the conventional one—only by 10%. This is due to low efficiency of utilization of natural uranium in such reactors—only the U-235 isotope is used whose content in natural uranium makes 0.72%, whereas the main component is U-238 (99.28%) whose fission probability in the TNR is very low. In addition, when current technologies are used, maintenance costs for radioactive waste and spent nuclear fuel (SNF) constantly increase. The society of the XXIst century will not be satisfied with increasing volumes of radioactive wastes of nuclear power stations, and simple refinement of the existing technologies will not give the required effect. No less acute problem is prevention of unsanctioned access to fissile materials and thereby strengthening their nonproliferation regime.

Therefore, the long-term strategy of the NPE development assumes transition to progressive technology, first of all, based on the use of fast neutron reactors (with energy of about 2 meV; so-called fast neutron reactors—FNR). It is essentially important that a much larger number of neutrons participating in the transformation of U-238 into fissile plutonium isotope Pu-239 be formed in the FNR in each nucleus fission act (burn up of natural uranium to 30–40%) for subsequent reprocessing of the fuel unloaded from the reactors of the NPP for subsequent reburying of unburned and newly formed fissile isotopes. The energy content of 100 g of uranium extracted from SNF is equivalent to that of 1 tons of oil, 2–4 tons of coal, or 1500–3000 m^3 gas. 19.5 thousand tons of new nuclear fuel can be obtained from 20 thousand tons of SNF. To obtain this amount of *fresh fuel*, it is necessary to mine and to process 6 million tons of uranium ore. Only at the expense of involving U-238 in the nuclear fuel cycle, the energy potential of the mined natural uranium can be increased by 100 times. As a whole, the use of "omnivorous" FRs in atomic power engineering opens a wide prospect for the creation of fuel for nuclear power stations in the form of artificial fissile elements for unlimited terms and transfers the nuclear fuel itself into the category of the renewable power resources.

Simultaneously with this, other positive effects are also reached:

Weapons-grade fissile materials are involved into the manufacture of electric and thermal energy,

Amount of SNF is reduced,

Thermal effect on the environment is reduced (due to a higher efficiency of nuclear power units);

The non-proliferation regime is strengthen due to the fact that the transportation of such materials is reduced to a minimum and the FR technology based on the closed fuel cycle is implemented in protective chambers under remote control with wide application of automation means. In this case, the third stage of the nuclear fuel cycle involves SNF delivery to the plant for its reprocessing and only after that final burial of waste of reprocessed FNR fuel.

4.5.4   Closed Nuclear Fuel Cycle

The *closed nuclear fuel cycle (closed NFC)* differs from the open one by the fact that after SNF holding in a temporal storehouse situated in the site of the nuclear power station, it is delivered to a radiochemical plant for reprocessing to extract the remained uranium (more than 95% of its initial mass) and built-up plutonium and manufacturing from them a new fuel. Simultaneously, radioactive isotopes of different chemical elements are separated and utilized. In addition, radioactive wastes are also separated. They are reprocessed and put inside the solidified glass mass buried in special burials. In this case, the NFC, including final burial of radioactive waste, lasts from 50 to 100 years.

However, despite the omnivorous FRs and the feasibility of implementation with their help of the closed NFC, they are not widespread because of a number of serious problems with their operation. Along with their seeming simplicity (absence of retarder and absorber), they are technically more complicated in comparison with TNR.

A number of serious problems must be solved associated mainly with the fact that the amount of fuel in the reactor required to obtain the chain reaction should be no less than a certain value called the critical mass. The FRs possess much larger critical mass compared to the TNRs (for preset dimensions of the reactor). In order that the FRs were no worse than the TNRs, it is necessary to raise the capacity developed for the preset dimensions of the reactor. To reduce the amount of fuel used per unit output, high density of heat generation should be provided. This leads to one more problem, because water as a coolant well mastered in thermal reactors does not suited for heat removal from the FR due to its nuclear properties. It slows down neutrons and hence, decreases the reproduction rate.

Fused sodium possessing good technological, thermophysical, and physical nuclear properties was chosen for heat removal from the FNR. It allowed high density of heat generation to be attained with reasonable measures of ensurance of high degree of safety. (The eutectic alloy of lead with bismuth is also considered as a coolant.)

There is one more special feature in the utilization of nuclear fuel in FRs. Under the influence of intense nuclear radiation, high temperature, and in particular, accumulated fission products the properties of the fuel composition (a mixture of fuel and raw materials) gradually deteriorate. The fuel forming critical mass becomes unsuitable for further utilization.

Due to the above-considered special features in FR operation, the cost of exploitation of nuclear power station built on FR basis (FNR-NPP) under projects of the 60s has appeared by a factor of 1.5–2 higher than of the NPP with TNR (TNR-NPS). For this reason, the FNR-NPP has lost the competition with the TR-NPP. The creation of large atomic power engineering (thousands of GW) planned already for the XXth century has been realized only partially and mainly based on TNR-NPP.

The fast-neutron reactors permitting to utilize up to 20 times larger portion of fissile material than in ordinary thermal-neutron boilers have already been built in Russia, the USA, England, and France. There are several breeder reactors with output powers in the range 250–350 MW all over the world. One of them has been operating in Shevchenko (now Aktau, Kazakhstan) since 1972. By decision of the government of independent Republic of Kazakhstan in 1999, its operation was stopped.

According to available predictions, achievement of competitiveness of the FNR-NPP (with reactors of the 4th generation) is expected only after 2025 with equivalent uranium price of the order of 200 dollars/kg.

In the USA the decision was accepted in the 70s to refuse from putting in operation breeder reactors and SNF reprocessing and to rely on the concept of the open NFC. At the same time, the project of an advanced modular safe (on principles of natural safety) fast reactor with sodium cooler and the project of improved FR with closed NFC for burnout of long-living actinides and fission products have been developed.

In Western Europe the leader in the FR development is France; however, the French program of the FNR-NPP development has practically been closed.

Prior to the accident at the NPPFukushima-1, *Japan,* unlike the United States and France, was developing a program for FR to sell them to other countries since 2020. However, Fukushima-1 NPP accident has compelled the government to declare freezing of works on the FR. (The project financing has been reduced by almost 80% in 2012.)

In India the demonstration sodium FNR (PFBR-500) is being developed and a small series of four FNR with capacity of 500 MW each is planned to be constructed.

The Russia is the leader in the development of FR of new generation. Exploitation in the USSR and then Russia of experimental-demonstration BR-5/10 reactor in Obninsk, BOR-60 in Dimitrovgrad, industrial BN-350 fast reactors in Shevchenko (nowadays Aktau, Kazakhstan), and BN-600 fast breeder reactor of the Beloyarsk Nuclear Power Station in Sverdlovsk Region (has been operating since 1980) during quarter of the century has proved the feasibility of the idea of regeneration of spent uranium, plutonium, and fission products to produce new fuel. (In the names of reactors is used the Russian abbreviation.) This has allowed Russia to start the development of the reactor of the 4th generation (FR) to proceed in perspective to the closed nuclear fuel cycle (the project "Proryv").

The FR with sodium cooler (BN-800) is under construction in the Beloyarsk NPP. The IAEA considers it as a perspective model for atomic power engineering of the XXIst century capable to provide in the near future the leading positions of Russia in this market. The experimental demonstration complex with nuclear cycle will be put in operation in 2020.

In Seversk (Tomsk Region) the construction of the FR with lead coolant is planned with implementation of the closed fuel cycle (Fig. 4.18) on its basis (Project "Brest-300").

Hopes with closure of the NFC are associated with putting in operation of this reactor based on burning weapon plutonium in MOX fuel. (MOX fuel is the mixed oxide uranium–plutonium (UO_2 + PuO_2) fuel.) Such fuel is planned to be manufactured approximately by 2015. By the same time, the FR series will be constructed. Modernization of the active zones will allow the most perspective variant of the reactor with reproduction rate of the new fuel sufficient for the developing APE to be chosen by 2030. Thus, the problem of utilization of weapon plutonium initiated works on the closure of the NFC based on the FR having the highest degree of technological readiness.

The FNR with sodium cooler (BN-800) was put in service at the end of 2015 in the Beloyarsk NPS. The IAEA considers it as a perspective model for atomic power engineering of the XXIst century capable to provide in the near future the leading positions of Russia in this market. The experimental demonstration complex with nuclear cycle will be put in operation in 2020.

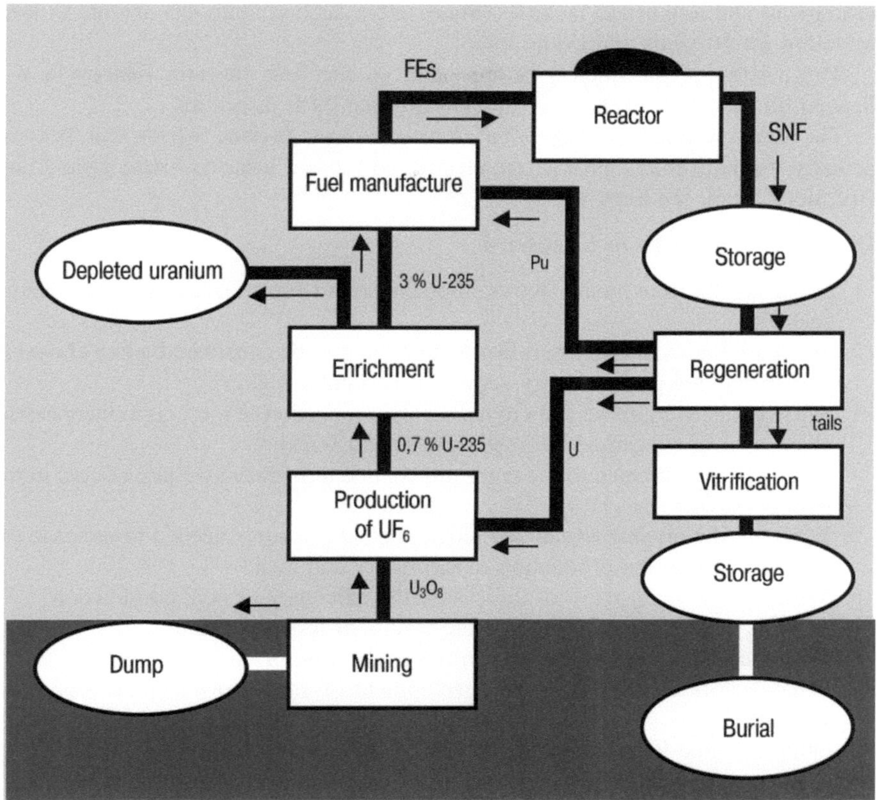

Fig. 4.18 Closed nuclear fuel cycle

The development and large-scale application of *a thorium fuel cycle* (with thorium-232 used as fuel) can be one of the ways of solving the problems of increasing nuclear fuel resource and reactor safety as well as of improving the ecological acceptability of nuclear power stations. The increased interest in this technology in the last few years has been largely caused by the fact that it guarantees better the regime of non-proliferation of weapon nuclear technologies. For this reason, fuel based on thorium and uranium is suggested for utilization of plutonium in reactors. Works on thorium cycle were carried out or are carried out in leading nuclear states (there are ready technologies and reactors in the USA and Germany, but they are conserved because of the high price of electric power generation) and in India and Brazil. In Russia low-intensive works are carried out in a number of scientific centers and universities. For example, the Tomsk Polytechnic University (Russia) developed the concept of atomic-hydrogen power station with reactor installations on thorium. The ultimate goal-creation of small power stations (about 10 MW) to ensure in future electric and thermal energy for remote settlements and production facilities (mines, etc.). Such systems can operate without refueling for 10 years and even more.

The destiny of the nuclear power plant on the base thorium fuel cycle will depend on successes in the development of other NPE directions.

The case in point is ***controlled thermonuclear fusion***, which can become actually inexhaustible source of the electric power and heat for future generations (for more detail, see Sect. 9.1).

Questions and Tasks to Chapter 4

1. What are the four main factors impeding rapid transformation in the energy sector?
2. What are the most important factors that have to be considered when choosing power plants type for energy supply of particular region?
3. Why do many countries plan to reduce the proportion of gas and to increase the share of coal consumed by power plants and boilers?
4. What dangers threaten to the environment due to increasing share of coal in the energy balance?
5. What are the organizational and technological measures needed to increase the share of coal in the production of electricity and heat?
6. What are the main ways of increasing the efficiency of coal combustion?
7. What is cogeneration? What are the results of its application?
8. Which countries are in the lead in the application of cogeneration?
9. What factors contribute to the development of small power (stand-alone and network)?
10. Why do some large companies prefer to have their small power plant?
11. What are the main types of installations of low power using traditional fuels?
12. What is the expander-generator?
13. What factors contribute to the spread of the gas turbine electric power generation technology in some countries?
14. Name the countries with the most developed nuclear power engineering.
15. What is the share of nuclear power in global electricity generation?
16. What is an open nuclear fuel cycle?
17. Advantages and areas of applications of small nuclear reactors, including on floating nuclear power plants.
18. The principles of operation of the fast reactors, their advantages and disadvantages.
19. What is a closed nuclear fuel cycle?
20. What is the fuel balance in power production?
21. Explain the possible reasons for the catastrophe in the nuclear power plant "Fucusima-1" and name its immediate and long-term consequences.
22. The impact of the radio-activity discovery, in the beginning of the XXth century, on the military sphere and power engineering.
23. What is the structure (without details) of a nuclear power plant?

24. What qualities of the power suppliers do you believe are necessary for the environmental protection?
25. What is the purpose of the floating nuclear power plant development?
26. What do you think about the prospects for the use of thorium as a fuel in nuclear reactors?

References

1. Warne DF. Newness electrical power engineer's handbook. House Elsevier; 2005.
2. Hunt S, Shutteworth G. Competition and choice in electricity. Chichester, England: Wiley; 1996.
3. Grigsby LL (ed). Electric power engineering handbook, 2nd ed. RCC Press; 2006.
4. El-Hawary ME. Electrical energy systems. CRC Press; 2000. p. 365.
5. Dugan RC, Mc Granaghan MF, Santoso S, Reaty HW. Electrical power systems quality, 2nd ed. 2004. World Development Indicators 07. Washington, DC: The World Bank; 2007.
6. Termuchlen H, Empsperger W. Clean and efficient coal fired power plants. New York: ASME Press; 2003.
7. EPRI 1000419, Engineering guide for integration of distributed generation and storage into power distribution systems. Palo Alto, CA: Electric Power Research Institute; 2000.
8. Momon J. Smart Grid: fundamentals of design and analysis, 1st ed. Wiley-IEEE Press; 2012.
9. Biomass for Energy and Industry. In: Proceedings of the international conference, Wurzburg, Germany, 8–11 June 1998.
10. Glachant J-M, Finon D, de Hauteclocque A. Competition, contracts and electricity markets. A new perspective. EE Publishing; 2011.

Chapter 5
Electric Power Engineering on the Basis of Renewable Energy Sources

5.1 Necessity of Searching for New Energy Sources

The primary reason for big popularity of the conversion of heat to electric power (about 70% of the total electricity production) has been the relatively low cost of fossil fuel sources (oil, coal, and natural gas) and the reasonably high system efficiencies that can be achieved owing to the high temperature differentials created by fossil-fuel combustion. Numerous sources indicate that the use of fossil fuels for power production must reduce to the level of sustainability, which means that alternative sustainable and environmentally responsible methods must be investigated and commercialized [1]. Evidence of such research and development can be found in the rapid growth of the international clean energy market from about $110 billion in 2006, to about $450 billion in 2016.

Among numerous factors determining the necessity of further gain in the power potential, despite a considerable increase in the efficiency of utilization of energy resources in a number of developed countries including Japan, Western and Northern European countries, and the USA, the following should be emphasized:

- increasing population (from 1.97 billion people in 1950 to more 7.0 billion people in first decade XXIst century),
- increasing electric energy consumption per capita (caused by the development and modernization of transport in connection with increasing mobility of population and growth of good traffics, a greater comfort of everyday life, further growth of urban population, relative increase in the share of country cottages, etc.), Fig. 1.4.
- further extension of inhabitant regions (1/3 of the Earth is not populated because of the absence of water, whereas half of the Earth's population endure a lack of space on the 1/10 of land); to this end, a cheap energy is required.

Due to the rapid growth of energy consumption and impossibility of its artificial limitation on the global scale, even in the face of the global threat of greenhouse

© Springer International Publishing AG 2018

V.Y. Ushakov, *Electrical Power Engineering*, Green Energy and Technology,

https://doi.org/10.1007/978-3-319-62301-6_5

effect and depletion of nonrenewable energy resources, we forced to solve immediately the following problems:

- develop power engineering based on utilization of unconventional renewable power sources (URPS),
- search for and master new methods of producing electric energy,
- improve the efficiency of utilization of energy resources with the use of energy- and resource-saving equipment, technology, and behavior of population.

Confirmation of the feasibility and high efficiency of these approaches can serve the economy of Denmark, Figs. 5.1 and 5.2.

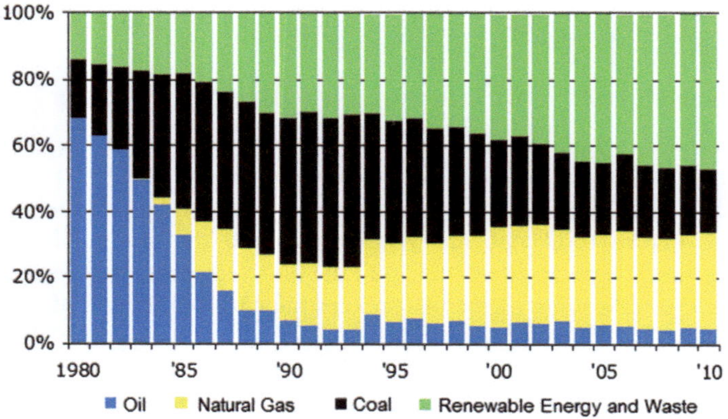

Fig. 5.1 Changes in the structure of energy consumption in Denmark in 1980–2010 and the result (Fig. 5.2)

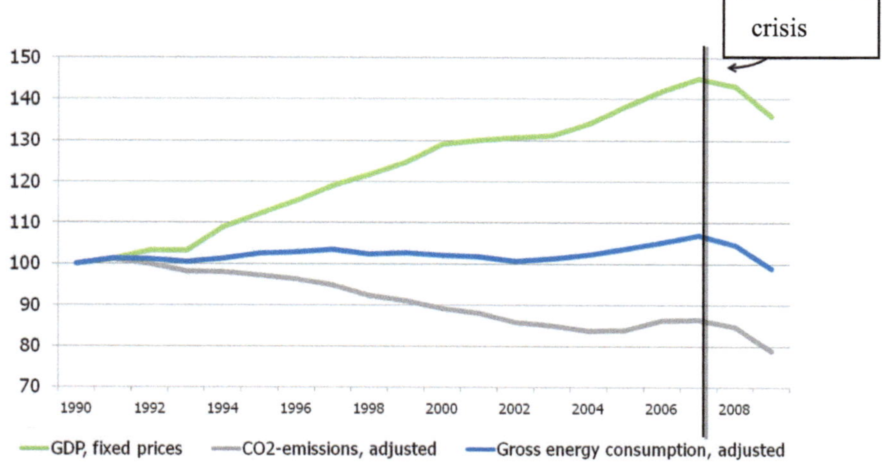

Fig. 5.2 Break the link between economic growth and energy consumption

Figures 5.1 and 5.2 illustrate the effectiveness of implementation of the principle of 3R in power engineering: **R**educing–**R**ecycling (**R**eusing)–**R**eplacing.

Successive energy policy of the Government of Denmark, which foresees more efficient use of energy resources and electricity, recycling of resources (the principle of "Zero Waste"), replacement of non-renewable sources of energy by renewable enables the country to develop rapidly in the time of simultaneous reducing energy consumption and emissions of greenhouse gases.

The term unconventional renewable power sources are used for solar, geothermal, wind, and tidal sources as well as for small river (the installed capability of units of hydroelectric power plants (HPP) varies from 0.1 to 10 MW), biomass, sea wave, and natural temperature gradient sources. As already indicated above, all these power sources are renewable.

At present, about 2 kW of electric power is generated per capita, whereas the life standard is 10 kW in the developed countries. (Annual consumption of electric power by each resident of a number of countries is shown in Table 5.1.) With allowance for the thermal barrier, the Earth's population must not exceed 10 billion people given that early or late the norm of power energy generation per capita

Table 5.1 Annual consumption of electric power by each resident of a number of countries

№	Country	Energy consumption (kWh)
	The whole world	*2933*
1	Australia	10,514
2	Belgium	8072
3	Brazil	2441
4	Canada	16,406
5	China	3312
6	Ethiopia	55
7	Finland	15,742
8	France	7318
9	Germany	7083
10	India	673
11	Israel	6927
12	Italy	5393
13	Japan	7847
14	Korea	10,162
15	Norway	23,174
16	Philippines	648
17	Russia	6533
18	Spain	5604
19	Sweden	14,029
20	Tanzania	92
21	United Kingdom	5518
22	USA	13,227

will reach 10 kW. Thus, the development of power engineering based on nonrenewable energy resources rigidly limits the population growth in our planet.

However, the Earth's population can reach more than 10 billion people to the middle XXIst century. Therefore, it follows that already now we must think about the reduction of the rate of population growth approximately twice; however, our civilization is not ready to do this. An approaching energetic-demographic crisis is obvious. This is one the most formidable argument for the development of un-conventional power engineering.

Many experts in power engineering consider that the only way out from this crisis is wide utilization of renewable power sources including solar, wind, oceanic, etc.

There is one more aspect of mastering URPS, especially important for Russia and some other countries having vast territories and low population densities, namely, power supply to customers. As already indicated above, the population of the most part of the vast territory of Russia (in the European North, Siberia, and Far East), about 12–15 million people, has no access to power supply from centralized power networks. They are supplied by electric energy mainly from autonomous diesel engines—low-power generators; fuel for them is delivered from distant regions using motor and water transport and even helicopters, which makes this fuel very expensive. Furthermore, these deliveries are unreliable, because they depend on weather conditions, availability of transport means, and prepayment.

Under these conditions, an alternative to organic fuel resources is the use of URPS in economically justified amounts. The humanity has made a great progress in the utilization of URPS in the last 15–20 years. As a result, due to large capital investments in this branch and legislative and political acts adopted by many countries on the intergovernmental level, the stage of development of systems based on URPS has progressed from research and development to industrial and commercial research. There are examples of elaboration and implementation of long-term programs. Thus, according to the Report of the European Commission, the EC activity in the field of URPS is aimed at coordination of efforts and large-scale utilization of URPS to ensure about 15% of energy generation in 2015 by utilization of RPS (including large-scale hydropower engineering). The global energy strategy of the Federal Government of Germany plans to increase the share of RPS taken together capacity to 50% by 2050. In addition, this plan is realistic, because only over the period of 1998–2002, it was increased from 4.6 to 7.1%. This result was reached only by increasing the competitiveness of units based on URPS compared to the conventional units. At present, as demonstrated below, conventional units surpass units based on URPS in the majority of technical specifications and commercial efficiency.

One of the most important characteristics of URPS—the specific capacity—is compared with conventional power sources in Table 5.2.

While on the subject of URPS, it should also be noted that many of them consume large amount of natural power resources per unit generated electric energy, Table 5.3.

Table 5.2 Specific power of unconventional renewable power sources

Source	Spec. power	Comments
The sun	100–250 W/m^2	
Wind	1500–5000 W/m^2	At a velocity of 8–12 m/s
Geothermal heat	0.06 W/m^2	
Wind-driven ocean waves	3000 W/m	Can reach 10,000 W/m
For comparison		
Internal-combustion engine	About 100 kW/L	
Turbojet engine	Up to 1 MW/L	
Nuclear reactor	Up to 1 MW/L	

Table 5.3 Renewable source energy consumed for electric energy generation

Type of the power system	Energy of a renewable source consumed per unit generated electric energy, rel. units[a]
Unit utilizing biomass	0.82–1.13
Heat- and-power station	0.08–0.37
Hydroelectric power stations	
Low-power	0.03–0.12
High-power	0.09–0.39
Solar photovoltaic units	
Ground-based	0.47
Orbital	0.11–0.48
Solar thermal unit (mirrors)	0.15–0.24
Tidal power station	0.07
Wind-turbine unit	0.06–1.92
Wave power unit	0.3–0.58

[a]Energy of a non-renewable source consumed per unit generated electric energy is 0.08–0.37 rel. units

Nevertheless, today considerably powerful power plants are developed which use renewable energy sources, Table 5.4.

Progress in the creation of reliable, technically perfect, and economically efficient electric power units simple in operation and based on utilization of unconventional renewable power sources will allow the main problem of decreasing the cost of a unit of generated power to be solved. From this viewpoint, of interest are the data of experts presented in Table 5.5.

Comparing conventional and unconventional power units and analyzing perspectives of their further development, we cannot but note their economic and ecological aspects presented in Tables 5.6 and 5.7.

As already indicated above, there are severe geographical, technical, and economic limitations on the application of NRPS.

The URPS resources have fixed geographical location. Obviously, tidal energy can be found at sea coasts, whereas the geothermal energy is generated by natural

Table 5.4 The most powerful power plants and individual power units, which use renewable energy

Type of power plant/power unit	Power (MW)	Cost (million dollar)	Country, firm-producer
Land wind power station (composition from 421 individual power units)	735		USA. Horse Hollow Wind Energy Centre
Offshore Wind Power Station	209	670	Denmark
Tidal power station ("Rance")	240	134	France
Tidal power unit ("Sea Gen")	1.2	6	Ireland
Solar power station (thermodynamic cycle)	392	2200	USA. Ivanpah Solar Electric Generating System
Solar power station (photovoltaic)		520	Spain
Geothermal power station	100		USA
Biomass power station	240 (elec.) + 160 (therm.)		Finland. Oy Alholmens Kraft
Wave power station (composition from 3 turbines/units)	2.25	1.3	Portugal. Aqcadoura Wave Farm

Table 5.5 Cost of power generated with utilization of different fuels and URPS, US dollars/kWh

Power sources	1980	1989	2000	2020
Non-conventional renewable power sources				
Solar energy	0.25	0.07	0.04	0.01
Thermal solar energy	0.24	0.12	0.05	0.03
Photovoltaic solar energy	1.15	0.35	0.06	0.02–0.03
Nuclear power plants and plants using an organic fuel				
Nuclear energy	0.04–0.13			
Energy produced by combustion of oil products	0.06			
Energy produced by combustion of coal	0.04			

springs of hot vapor and water and thermal anomalies. Small HPS, naturally, can be built on rivers and storage reservoirs. Solar and wind energies are widespread everywhere, but their utilization is most expedient in regions with maximum solar irradiation and maximum wind velocities, respectively.

Technical limitations are caused by the problem of stability of power networks and individual power units with uncontrollable power generation charts. The problem of the power networks stability is particularly strong when the power produced from renewable energy sources is approaching or exceeds 20–25% of the power networks power. The increase in power generating units that use URPS complicates the task of storing electrical energy, as well as the conversion of generated direct current into an alternating to increase voltage level up to the

Table 5.6 Material and labor consumption for building and operation of electric power units of the indicated types

Primary energy resource, power source	Material consumption of the power unit, rel. units	Total labor consumption for building and operation of the power unit, rel. units
Natural gas	1.0	1.0
Oil	2.2	1.6
Coal	3.2	2.0
Nuclear power engineering	5.6	2.8
Solar energy		
Heating	62.5	40.0
Photo-conversion	109.4	140.0
Hydraulic power	62.5	–
Wind energy	250.0	72.0

Table 5.7 Average floor area, in m^2, required for annual production of 1-MW electric energy on power plants of the indicated type

Type of power plant	Floor area (m^2)
Nuclear power plants	630
Heat-and-power plants using	
Liquid fuel	870
Natural gas	1500
Coal	2400
Solar power plants	100,000
Hydroelectric power plants	265,000
Wind power plants	1,700,000

working voltage of power networks. The second problem relates to the photo-electric converters, fuel cells, some types of WPP. (Briefly, they will be discussed in Sects. 5.4, 5.5, and 9.2.)

Technical limitations are caused by the problem of stability of power networks and individual power units with uncontrollable power generation charts.

The economic efficiency of power stations is also very important now. The majority of power stations utilizing URPS are not competitive with traditional ones. This is due to low flux density (specific power) of utilized primary energy (Table 5.2) that leads to material and capital expenditures for systems utilizing URPS higher than those for systems utilizing conventional energy resources (Tables 5.6 and 5.7). If expenditures on a power station utilizing URPS are covered in acceptable time, its initial cost is almost without exception higher than that of a conventional station, which creates a barrier for URPS utilization; therefore, the technical policy of many countries in the given field provides economic incentives to the NRPS development consisting in tax and credit privileges or granting

Table 5.8 The share of the NRPG in the world Fuel and Power Complex

Country, continent	Share of NRPG (%)
Africa	0.09
Asia	1.4
Latin America	2.7
Australia	1.85
EU Countries as whole	2.7
Denmark	12.3
Germany	3.0
Spain	2.9
United Kingdom	2.0
Russia	<1.04 (4.5% in 2020)

subsidies to the manufacturers and customers of equipment. In these countries, the appropriate legislative and normative base has been adopted according to which, in particular, energy companies are obliged to purchase the electric energy produced by utilizing URPS. These measures are necessary for the NRPS development in the initial stage in which power generating units and stations based on NRPS can and must become competitive with the conventional ones.

As noted above, in most countries the share of the non-traditional renewable power generation (NRPG) in the world Fuel and Power Complex is too small, Table 5.8. The share of different types of NRPG in the total volume of using NRPG globally and in the leading countries and associations of countries is shown in Fig. 5.3.

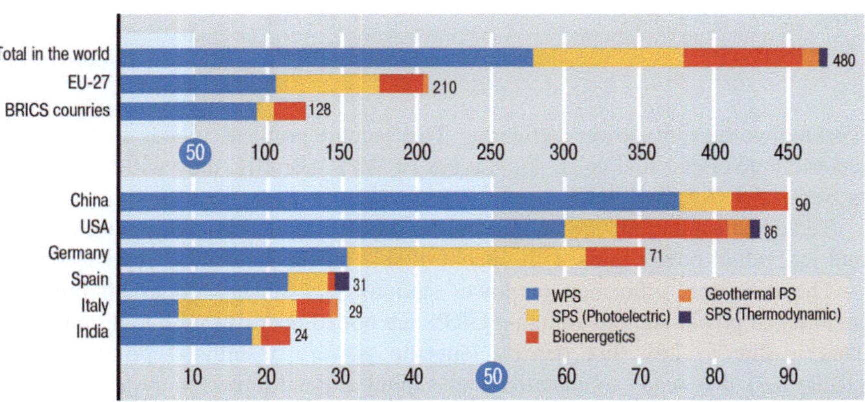

Fig. 5.3 Capacity of generating installations using different types NRPG globally and in the leading countries and associations of countries

Table 5.9 Waterpower resources in the indicated countries

Countries	Power (GW)		Countries	Power (GW)	
	For average annual water discharges (50% coverage)	For minimum water discharges (95% coverage)		For average annual water discharges (50% coverage)	For minimum water discharges (95% coverage)
Russia and former republics of the USSR	249.4	79.5	France	5.8	3.4
The USA	53.9	25.0	Italy	5.2	2.8
Canada	25.1	15.85	Switzerland	3.8	2.4
Japan	13.2	5.6	Spain	5.0	2.9
Norway	20.0	12.0	Germany	3.7	1.5
Sweden	8.9	2.9	England	1.2	0.6

5.2 Harnessing of Water Flow of Rivers and Energy of Other Streams

In contrast with non-renewable chemical energy accumulated in an organic fuel, the kinetic energy of river water is renewable. It is transformed into electric energy at hydroelectric power plants.

The estimated annual waterpower resources on the Earth are 32,900 TWh, from which only about 25% can be utilized for technical and economic reasons. This value is nearly twice as large as the current level of annual electric power generation by electric power plants all over the world, that is, definite reserves of waterpower are still available. Table 5.9 presents the data on waterpower resources in different countries.

5.2.1 Large-Scale Hydraulic Power Engineering (Base on Traditional Hydroelectric Power Plants)

In 1882, the first water wheel-driven generator was installed in Appleton, Wisconsin (USA). The low voltage of the circuits limited the service area of a central station, and consequently, central stations proliferated throughout metropolitan areas.

Nowadays power plants converting the energy of river water into the electric energy provide the basis for the part of power engineering based on renewable power sources. For almost centenarian history of the development of hydraulic power engineering, huge experience on the hydraulic power plants (HPP) construction on mountain streams and flat rivers, on rivers with giant water discharge and on small rivers (hydroelectric power micro-plants) has been accumulated together with the experience on building of fluvial and dam HPP, and hydraulic power engineering itself is among the most significant achievements of the XXth century. It is commonly accepted to identify four types of hydraulic power systems:

- large HPP that form hydro-engineering complexes,
- pumped storage power plants,
- small HPP operating outside of interconnected power systems,
- hydroelectric power generating units being integral parts of national-economic complexes.

Currently, the bulk of electricity generated by a water flow accounted for HPP is built on large rivers. They have been built according to one of the three schemes: dams, derivative, or combination. The typical design of the hydroelectric unit ("heart") of the powerful HPP is shown in Fig. 5.4.

At present, about 30% of the world economic hydraulic potential is utilized.

The relative contribution of hydroelectric power to the total power generation is rather high in many countries: in Norway and Brazil, it exceeds 90%, in Canada and Venezuela it is between 50–80%, in India, Egypt, Italy, and China it is about 20%.

The hydropower potential of the rivers in different countries and its implementation are shown in Table 5.10.

The indisputable advantages of hydraulic power engineering are:

- renewable water-power resources,

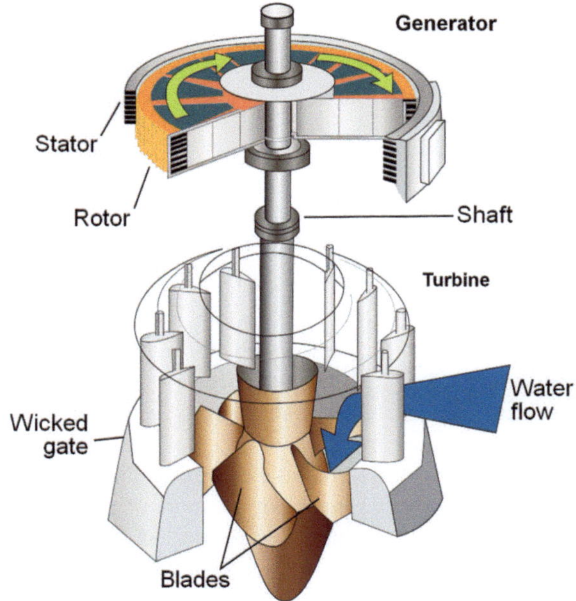

Fig. 5.4 Typical design of the hydroelectric unit of the powerful HPP

Table 5.10 Specifications of the world HPP

Country	Installed capacity (MW)	HPP output (GWh/year)	Economically justified hydroelectric potential (GWh/year)	Number of HPP	Percentage of hydroelectric capacity in total output (%)
USA	75,525	308,800	3,760,000	6389	8.8
China	72,900	212,900	1,260,000	24,119	17.3
Canada	65,726	350,000	536,000	820	62.0
Brazil	56,481	301,198	736,500	470	93.5
Russia	43,940	157,500	852,000	62	16.4
Norway	27,410	116,259	176,600	300	99.4
France	23,100	69,800	71,500	554	15.0
India	21,963	74,338	No data	2601	25.0
Japan	21,389	102,587	114,267	2467	10.0
Spain	17,000	39,000	41,600	871	20.0
Sweden	16,204	69,300	90,000	144	47.7
Italy	15,267	51,636	54,000	502	19.4
Venezuela	13,224	57,923	100,000	69	73.0
Turkey	10,215	42,229	123,040	427	38.0
Mexico	9702	24,616	32,232	540	14.4

- low production cost of electric energy (The average cost of the electric power produced by HPP in Russia is almost 6 times less than that of the electric power produced by TPP state district power plant despite of the fact that domestic prices for fossil fuels in Russia are relatively low),
- the mobility of reserve capacity covering peaking load charts,
- inflation stability,
- ecologically clear production,
- the ability to build a giant power HPP, Table 5.11.

The accumulated knowledge allows us to pay more attention to the ecological factors now, in particular, to positive effects that can be obtained by constructing objects of hydropower engineering. A comprehensive analysis enables us to evaluate the effect of HPP on the environment.

The pace of development of powerful hydropower is globally limited by the following factors:

1. In some countries, there are no rivers with sufficient hydropower potential (e.g., Israel).
2. In some countries it has been involved practically all (e.g., Norway).
3. Many countries with large hydroelectric potential have no money to invest in construction of such facilities. As is well known, the construction of powerful hydroelectric power station is characterized by extremely high capital expenditures (billions of dollars) and long construction schedule (up to 10 years or more);

Table 5.11 The most powerful HPP in the world

№	Name of HPP	Country	Starting year	Full power (GW)	Annual production	Area of reservoir
1	Three Gorges	China	2008	22.5	More then 100	
2	Itaipu	Brazil and Paraguay	1984	14	94.7	1.35
3	Guri	Venezuela	1986	10.2	46	4.25
4	Tucurui	Brazil	1984	8.37	21	3.014
5	Grand Koulee	USA	1942	6.809	20	
6	Sayano-Shushenskaya	Russia	1975	6.4	26.8	0.621
7	Krasnoyarsk	Russia	1982	6	20.04	2
8	Robert-Bourassa	Canada	1981	5.616		
9	Churchil Falls	Canada	1971	5.429	35	6.988
10	Longtan Dam	China	2009	6.3	18.7	
11	Bratsk	Russia	1967	4.5	22.6	
12	Ustilimsk	Russia	1980	4.32	21.7	

Note The table shows the year of start up of the first unit operation. The appearance of the world's largest hydropower plant is shown in Fig. 5.5.

Fig. 5.5 Photograph of the HPP Three Gorges (China)

4. Despite the apparent ecological purity of production of electricity from hydro-power plants, they have a significant negative impact on the environment (see Chap. 8).

More detailed data on the history and state of the art of large-size hydro-power engineering can be found in the special literature.

5.2.2 Mini Hydro Power Plants

Except the above-mentioned division of HPP by their functionality, the following classification of HPP by their capacity is used:

- stations with capacities smaller than or equal to 100 kW are called *micro* HPP,
- from 100 to 1000 kW are called *mini* HPP,
- from 1000 to 10,000 kW are called *small* HPP,
- more than 10,000 kW are called *large* HPP.

Their design and principles of their construction can differ essentially.

Historically, the first HPP belonged to the class of *micro* HPP, and the time of their appearance coincided with successes in industrial mastering of electric generators. These simplest often half-handicraft systems were widespread, especially in rural regions. In particular, the share of hydropower engineering in energy supply in agriculture reached 11% in the USSR in 1937.

In the USSR, in contrast with the majority of other countries where hydraulic power engineering was developed simultaneously with other power resources, it began virtually from the very beginning. When small power engineering was ignored in general and *micro* HPP were ignored in particular, even the experience of utilization of small rivers was lost, many operating hydroelectric systems were closed, and the production of equipment for them was cut down.

Due to achievements in the field of electric machine building and semiconductor and transformer engineering, a new class of valve-type electric machines, which possess new properties and allow one to solve problems previously unsolvable, has been developed.

For example, the valve-type electric machines provide the basis for the construction of autonomous electric power supply units that generate high-quality electric power with minimum requirements imposed on an actuating motor. This enables one to construct automated hydraulic unit with uncontrollable turbines for hydraulic power micro-stations. As demonstrated the pre-war experience, exactly this direction in the development of *micro* HPP meets best the production-technological and operating requirements. *Micro* HPP with valve-type electric machines are most widespread now all over the world.

The tendency to a simplification of hydraulic engineering equipment of stations increased strongly the requirements to devices for electric power generation and stabilization of its parameters.

The main directions in the development of small-scale hydraulic power engineering in Russia for the nearest future are the following: construction of *small* HPP on complex water engineering systems; modernization and reconstruction of outdated *small* HPP; construction of small HPP on small rivers; construction of small HPP on existing water basins, water falls of canals, penstocks for water supply to and removal from various economic objects.

In 2014 in the Tomsk region (Russia) a mini HPP using sewage waters of two cities (Tomsk and Seversk), settlement and petrochemical plant was put into operation. Its parameters are power of 1000 kW, height difference of 96 m, water pipeline diameter of 1420 mm, and the rate of water discharge 3500–11,000 m^3/h, and the payback period of 3 years.

5.3   Bioenergetics

Bioenergetics can be considered as a variant of solar power engineering (see Sect. 5.5) based on photosynthesis and subsequent liberation of chemical energy accumulated in biomass, which is converted into thermal or electric energy. The biomass is the cheapest and largest-scale form of renewable energy accumulation. The term **biomass** involves any material of biological origin, including products of vital activity and organic wastes. A special place is occupied by *peat*—one of the materials of biological origin. (Sometimes it is called the *accumulated biofuels*.) Peat is often referred to non-renewable energy resources due to the very low rate of recovery of its reserves.

The average rate of peat accumulation varies and depends on the prevailing source of plant grouping, geographical and climatic zones, and hydrological and other conditions. It varies from 0.2 to 0.4 mm (swamp forest—tundra) to 1–2 mm (coniferous-deciduous subzone). Modern peat deposits were formed 10–12 thousand years ago.

5.3.1   Biomass

One of the most complicated problems on the paths to implementation of a tempting idea of plant utilization as a main energy source is a low efficiency of photosynthesis as a method of converting solar energy into chemical one. According to estimates, about 155 billion ton of dry residue of organic mass, mainly cellulose, which can be used either directly as a fuel or as a material for producing fuel, is formed annually in the process of photosynthesis. Because of a low efficiency of energy conversion, areas under crop must be increased significantly to produce the required amount of energy. Therefore, investigations aimed at an increase in the efficiency of energy conversion, a search for plants most suitable for this purpose, and creation of an optimal artificial gas composition are of great importance. For example, if corn is grown to produce energy rather than to feed cattle, the cost of energy will be comparable with the cost of fossil fuel in the USA. If coniferous forest is grown with 6 thousand trees per 1 acre (1 acre = 0.4 ha), and a "harvest" is gathered every 12 years, the cost of energy produced from it will increase approximately twice and will be $\sim$ 3 dollars per 1 million British Thermal Units (1 Btu = 1.05506 $\times$ 10^3 J $\approx$ 1.055 kJ) due to slow rate of growth of trees and some other factors. Perennial plants have one inestimable advantage over annual ones: the "harvest" can be gathered all the year round when required; in this case, we need not construct huge granaries for "energy harvest" collected only in definite season. Therefore, fast growing deciduous trees, whose roots produce new sprouts after cutting thereby obviating the necessity of new annual planting, were chosen to produce energy.

Hybrid poplars are grown in experimental sites of unused arable lands in Central Pennsylvania. Each of 3700 trees of a hybrid planted on an acre produces energy that costs from 1.25 to 11.45 dollars per 1 million Btu (compared to a cost of 1.97 dollars per 1 million Btu for oil and 1.31 dollars per 1 million Btu for coal). This plantation can produce annually $\sim$120 million Btu per acre with a power conversion efficiency of $\sim$0.6%. To supply with fuel an average electric power station rated at 400 MW, the plantation with an area of 30 thousand acres is required. To fuel all electric power stations of the USA, a plantation with an area of no more than 160 million acres is required even if the coefficient of solar energy conversion into fuel is $\leq 0.4\%$.

In the present stage, the annual increment of organic matter on the Earth is about 170–200 billion tons of biomass in the re-run of dry organic matter, which is equivalent to 70–80 billion tons of oil (4 $\times$ 10^{21} J). This is approximately 10 times more than global total energy commercial consumption of all humanity (3.9 $\times$ 10^{20} J), Table 5.12.

Biomass sources can be subdivided into the following main groups:

1. Products of natural vegetation (wood, waste products of wood working, leaves, etc.).
2. By-products of people activity including industrial activity (solid domestic wastes, wastes of industrial production, etc.).
3. By-products of agriculture (manure, chicken dung, stems, tops, etc.).
4. Specially grown high-yield agricultural crops and plants.

However, the biomass itself even in large amounts does not imply a solution of the problem of obtaining different products and substances from it, including fuel. The unutilized biomass does irreparable harm to the environment.

By-products of woodworking have already been utilized: facilities have been created and technologies of production of a generator gas and of its combustion are mastered. The experts believe that 15% of need in fuel can be covered by rational utilization of wood, by-products of wood working, and fast growing forest plantations. With the modern level of consumption, this will make about 6 million ton of standard fuel.

Table 5.12 Biomass sources and examples of biomass utilization

Biomass source	Biofuel	Technology	Estimated conversion efficiency (%)	Need for energy ("n" stands for necessary, and "o" stands for optimal)	Estimated output power of biofuel (MJ/kg)
Logging areas	Heat	Combustion	70	Drying (o)	16–20
Waste products of wood working (a)	Heat	–	70	Drying (o)	16–20
(b)	Gas Oil Coal	Pyrolysis	85	Drying (o)	40[a] 40 20
Grain	Straw	Combustion	70	Drying (o)	14–16[c]
Sugar cane, juice	Ethanol	Fermentation	80	Heat (n) Electric power (o)	3–6
The same, waste products	Oilcake	Combustion	65	Drying (o)	5–8
Manure (the tropics)	Methane	Anaerobic decomposition	50	–	4–8[c]
The same (temperate zone)	Methane	The same	50	Heat (n)	2–4[b]
Domestic sewage	Methane	The same	50	Heat	2–4[c]
Garbage	Heat	Combustion	50	–	15–16[c]

[a]Total value; biogas is consumed for heating of the unit
[b]Disregarding nitrogen
[c]Dry residue

At present more than 6% of all thermal energy consumed in China, nearly 6% in the USA, 5.7% in the EU countries, and 32.9% in Brazil are produced from biomass. In the countries of the equatorial belt, biomass remains the main source of energy. Its share in the energy balance of developing countries is 35%. The one sixth of annual fuel consumption in the world is provided by wood and about one third of all cut-down trees are used for cooking and heating. About half the world population uses firewood for cooking and heating (and this is 4/5 of energy consumption).

Developed and applied various methods of converting biomass into thermal and electrical energy or liquid and gaseous fuels, Fig. 5.6.

It is possible to distinguish three main methods of producing fuels from biomass.

First, bioconversion is used, that is, decomposition of organic substances of vegetative and animal origin under anaerobic (without access of air) conditions by special bacteria with the production of gaseous fuel (biogas) and/or liquid fuel (ethanol, butanol, etc.). Now in Brazil the urban motor transport and many personal automobiles work on fuel produced by decomposition of biomass from sugar cane.

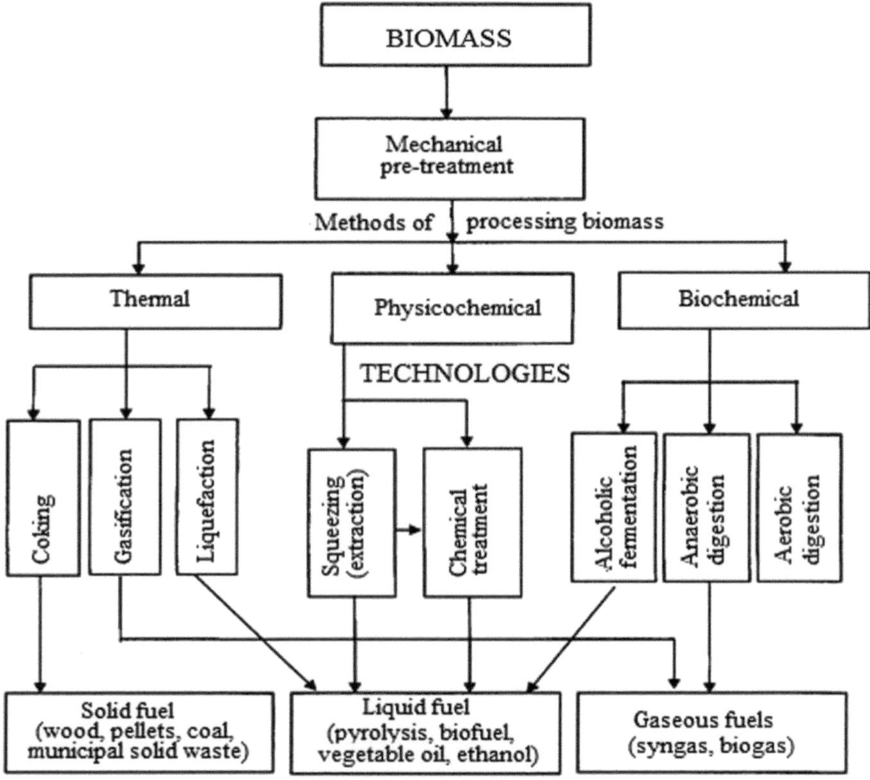

Fig. 5.6 Methods of processing technology and the use of biomass energy

In the USA, ethanol is produced from wastes of corn. Ethanol is a good substitute for gasoline; moreover, biomass is a fast renewable resource unlike oil. The production of thermal energy through aerobic microbiologic oxidation of organic substances also belongs to bioconversion. This is the scientific term for composting and bio-heating known to each gardener.

Second, thermochemical conversion is used (pyrolysis, gasification, fast pyrolysis, and synthesis) of solid organic substances (tree, peat, and coal) into a synthetic gas, methanol, artificial gasoline, and charcoal.

Third, combustion of wastes in special boilers and furnaces is used. Hundred tons of such wastes are burnt with regeneration of energy all over the world. The calorific values of pressed fuel bricks from paper, cardboard, wood, polymers, sawdust, and garbage are comparable with the calorific value of brown coal.

The use of **secondary resources**—agricultural waste and wood processing, combustible household waste and industrial waste has obvious advantages over the use of primary biological resources (primarily wood, food, and feed crops):

- does not threaten the biological balance;
- improves the environment;
- does not lead to higher prices for food products as is the case of the large-scale use of the food and feed grains in bioenergetics.

In developed countries, each person produces about 5 tons of dry organic waste per year, processing of which in the methane can give in a global scale to 700 billion m^3 of methane or 1×10^9 tons standard fuel. This is sufficient to meet the needs of all population in the gas for cooking. Such processing can help to solve environmental problems and to produce fertilizers.

In developed countries, to generate heat and electricity by direct burning of firewood, waste wood, straw, and peat mini-power plant (Mini-TPP) is used. In the embodiment of steam turbine, their efficiency is 20–25%, power from a few kilowatts (for farmers) to hundreds of kilowatts, Table 5.13.

Increased efficiency and lower operating expenses have automatic boilers that require pre-treatment before combustion of biomass: the manufacture of wood pellets, wood chips, and briquettes.

Pellets are 20–50 mm granules from dried wood waste manufactured under high pressure without chemical fixers; their cost is 60–90 Euro/t, and cost of unit energy produced from these is approximately 30% lower than that of diesel fuel or natural gas.

Table 5.13 Key characteristics Mini-TPP

Power (elec.) (kW)	Power (heat.) (kW)	Fuel expenses (kg/h)	Weight (kg)
5	200	80	1500
16	610	250	5000
30	1200	400	8000
120	1500	800	15,000
200	2800	1400	28,000

Table 5.14 Comparative characteristics of pellets and other fuels

Type of fuel	Calorific value (MJ/kg, MJ/m^3)	Sulfur (%)	Ash (%)	CO_2 (kg/GJ)
Diesel fuel	42.5	0.2	1	78
Fuel oil	42	1.2	1.5	78
Natural gas	35–38	0	0	57
Coal	15–25	1–3	10–35	60
Wood pellets	17.5	0.1	1	0
Straw pellets	14.5	0.2	4	0
Peat pellets	10	0	4–20	70
Wood chips	10	0	1	0
Sawdust	10	0	1	0

Fuel pellet characteristics and some other types of fuel are shown in Table 5.14.

Due to the advantages of pellets, they become one of the most popular types of fuel in most European countries, Japan, the United States and in several other countries. In Europe, annual growth in the production of pellets is about 30%.

The main consumers of the pellets are the owners of country houses that use this fuel in fireplaces and mini-boiler.

Energy chips have a length of 10–150 mm and a thickness of 10 to 100 mm. Often chips are burned with sawdust, which allows one to recycle waste and to improve fuel consumption by about 20%.

Briquettes are made from timber waste, straw, tops, etc. Their sizes and shapes are different: rectangular, cylindrical; size—from meters to centimeters (depending on the type of boiler and BM). Their calorific value is close to that for coal, Table 5.15.

To reduce emissions of harmful substances, to reduce the requirements for preliminary fuel preparation, and to improve the efficiency of fuel conversion, the following biomass combustion technologies are developed and used: combustion in swirling furnace (cyclone combustion chamber), and use of the powdered fuel, grill, rotary kiln, fluidized bed, and circulating fluidized bed. Each of them has advantages and disadvantages. Two latest technologies are the most preferable from the point of view of ecological cleanness. Boilers with low-temperature furnaces and fluidized bed allow biomass with humidity of 60% or more to be burned and the swirling furnace (cyclone combustion chamber) allows pulverized wood and plant waste to be burned.

Table 5.15 Calorific value of fuel of different types

Type of fuel	Calorific value (kcal/kg)
Wood (wet)	2450
Wood (dry)	2930
Brown coal	3910
Briquettes from waste wood	4400
Black coal	4900

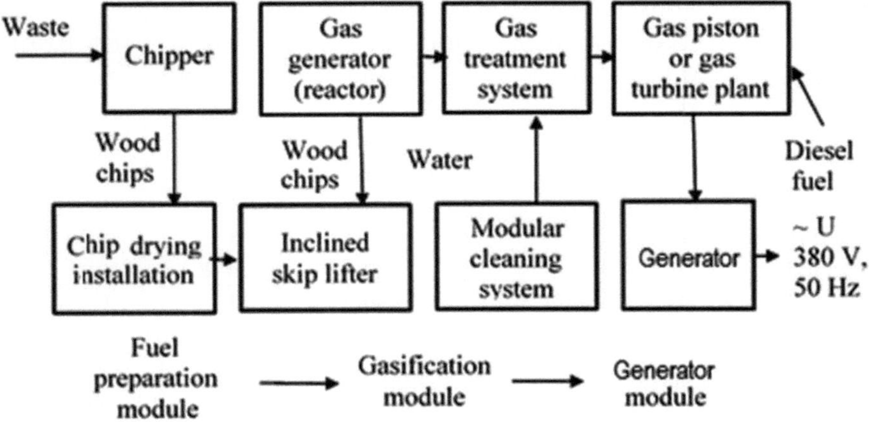

Fig. 5.7 Block diagram of the gas-producing power plant

Table 5.16 Characteristics of the gas-producing power plants using wood and wood waste

Power (kW)	Fuel consumption (kg/h)	Weight (kg)
8	20	1000
12	24	1500
30	52	1800
70	120	2500
100	170	3000
200	380	3500
500	800	6000

Increase in the efficiency of the conversion of biomass into heat and electricity by more than 2 times compared to TPP is possible due to preliminary biomass gasification in the gasification modules included in the equipment of set piston and gas turbine power plants (it can also be used in TPP with steam boilers). Scheme of gas-producing power plant is shown in Fig. 5.7; main power characteristics of TPP of this type are shown in Table 5.16.

5.3.2   Peat

The reserves of one more energy source—*peat*—are also significant.

World peat reserves are estimated at more than 500 billion tons (by a 25% moisture content, they amount to 225–261 billion tons). The area of peat deposits in the world accounts for 176 million hectares. Deposits of peat are found on all continents. A relatively small part of the Earth's land surface (in the zone of maximum concentration of peat) contains more than 80% of world reserves of peat.

In the Northern Hemisphere, it covers Western Siberia and extends to West Atlantic coast. A large area of peat accumulation is in the northeastern part of North America. In the Southern Hemisphere, significant deposits of peat are found only on the islands of Southeast Asia.

According to estimates of the Canadian Peat Resources (2010), the first place in the world reserves of peat take Canada (170 billion tons), the second—Russia (150 billion tons), the third—Indonesia. The annual increase of peat reserves in Russia is 250 million tons, which is 50 times higher than the level of annual procurement.

According to the share of peat in energy production, the leader is Ireland, where more than 90% of power plants work on peat.

The construction of the first power plant using peat as fuel (peat power plant) began in the late 1870s–early 1880s. (In Russia the first peat power plant was built in 1913, today it is the largest in the world TPP (with power of 1500 MW). In Russia, the largest plant that can work on peat is Shatura SDPP.

In Finland and Sweden, the power plants working on peat produce a quarter of total produced electricity. In Finland, peat is used for about 60 TPP owned by industrial enterprises and municipalities. The power of boilers is 20–550 MW, and the total capacity of these TPP is more than 7 GW.

In assessing the competitiveness of peat with other fuels, it is conveniently compared with coal. In this case, heat has the following advantages:

- the replacement of coal by peat significantly reduced air pollution by sulfur oxides (by 4–24 times depending on the ash content and the characteristic coal basin),
- by 2–19 times reduced emissions of particulate matter,
- utilization of peat ash is simple compared to the utilization of coal slag,
- the development of peat deposits reduces the risk of fires.

Because of low calorific value, peat has not yet found wide application in electric power engineering [2].

Low calorific value of peat causes many problems when it is used in the power industry: at current prices for mineral fuels, the peat share in the energy balance is very small.

5.4 Wind Power Plants

The wind energy has long been used in navigation and for turning round mill wheels too. In recent years, it has been utilized for electric power generation. The majority of wind power units have a capacity of a few kW, and they are located in distant regions, for example, on the sea shallows, the so-called offshore wind farms. During the Second World War, a wind power plant rated at 1.25 MW was constructed on the Grandpa Hill, Vermont, USA, which had been operated successfully for several weeks and generated 61.78 MWh of electric power. Then one of the rotor blades was broken, and the station was not restored most likely because of a deficit of materials and the need for saving money during the time of war.

Since the energy crisis in 1973–1974, considerable means have been invested in the development of wind power engineering. Several experimental stations of different designs were constructed. The cost of the electric energy generated by wind power stations is still higher than that generated by power stations on organic fuel (see Table 5.5). In addition, some more disadvantages peculiar to this source of primary energy were found. Nevertheless, the wind energy should be considered as an important energy resource.

It can be easily shown that the unit output power is proportional to the area of wind rotor blades and to the cube of wind velocity (usually low). In this regard and due to the fact that the density of air is 846 times less than the density of water, the overall dimensions of wind high-power plants (in megawatt range) should be very large compared to hydraulic turbines (Fig. 5.8).

One of the most complex problems hindering widespread use of wind power plants is the permanently varying wind velocity. Even in high mountains, the wind velocity undergoes fast variations.

As the rotor speed varies significantly, a gearbox is installed to regulate the speed of the shaft on which the power generator is mounted. A DC-generator is an easy-to-control system for converting the mechanical energy to electric power.

Fig. 5.8 Photo of a 4.5 MW windPP with a diameter of 112.8 m and a gondola at a height of 180 m above the ground

However, problems associated with the type, quality, and transmission of the power must be solved in that case. The energy efficiency of WPP may be significantly increased based on a synchronous generator with high-power electronic equipment. In that case, the voltage of the synchronous generator is sent to the input of rectifier, which may be based on uncontrollable switches in some circumstances. Then, rectified voltage is sent to an inverter (Fig. 5.9), which is connected directly to the grid. Either a voltage inverter or a current inverter may be used. Its operation with the grid can by greatly improved by using completely controllable electronic switches. By that means, the inverter may generate specified active and reactive power in the grid. It is also possible to insure high quality of the current and to supply energy from the grid for battery charging.

Another option is to use an induction generator (IG) with a short-circuited rotor, but that entails regulation of the amplitude and frequency with variation in the rotor speed. Special methods are required here, for example, combined operation with a rectifier-inverter cascade. An inverter based on completely controllable electronic switches mast be used in that case; a step-up DC-converter may also be needed. However, a more promising approach is to use a source of controllable capacitive power such as static converter. In that case, the IG can send active energy to the grid and draw from it the required reactive power, Fig. 5.10.

A fundamentally new approach involves an asynchronized synchronous generator. Its rotor has two windings, which may effectively be controlled by static converters. In principle, this generator may operate with the grid as its rotor speed varies. Power may be supplied to the rotor windings from a frequency converter in which the voltage is phase-shifted by 90°, and its frequency is the slipping frequency. The resulting magnetic field, which turns at the slipping frequency, ensures synchronous rotation of the exciting field and the generators stator. The efficiency of the asynchronized synchronous generator is especially evident at considerable turbine speeds with a gearbox of limited capabilities. The frequency control of the voltage supplied to the winding ensures highly stable synchronous operation of the asynchronized synchronous generator with the grid [3].

In addition, these plants generate electric power only when the wind blows rather than when required. Unfortunately, there is no convenient, efficient, and economic

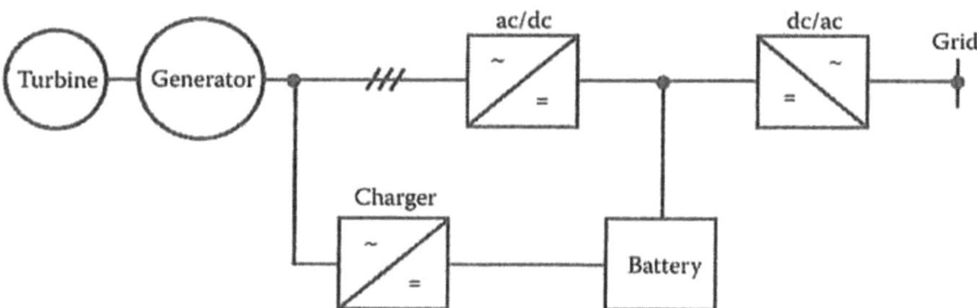

Fig. 5.9 Block diagram of a wind power system

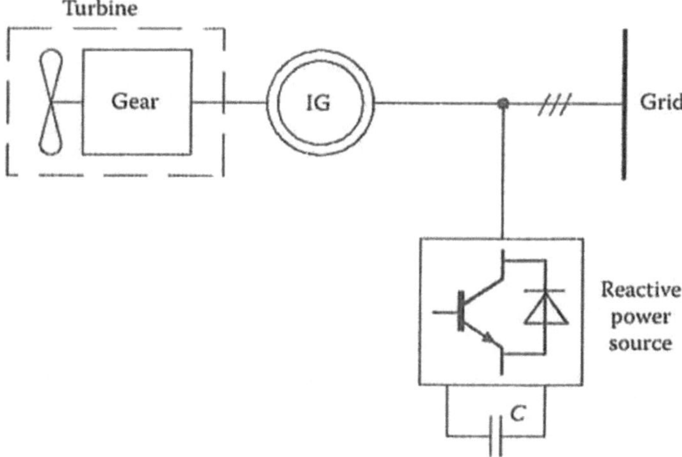

Fig. 5.10 Connection of wind turbine to the grid by means of an IG

method of electric power accumulation in large amounts. In addition to the principle implemented in WPP, proposals on utilization of electric power generated during periods of minimum electrical loads for electrolysis of water and production of hydrogen and oxygen are nominated. These gases can subsequently take part in the association reaction in a fuel cell to generate electric power. The development of this technology has only been started. However, it seems likely that in future it can become economically acceptable.

In the early 2002, the beginning of implementation of an almost fantastic project was reported in press. According to this project, the synthesis of solar and wind power engineering is envisaged, independent from weather conditions, time of day, and season.

The development of wind power engineering in the world began in the early 70s.

The greatest share (up to 3%) of the WPP generation was recorded in 1993 in Denmark with wind turbines distributed all over the country. The construction of modern WPS was started here in the late 70s. In the early 80s, especially rapid growth of WPS was observed in California, USA. The law on tax discounts for investments to renewable power sources, adopted in addition to federal tax discount law, has created a favorable situation here. As a result, California is now the world leader in production of wind electric power. The most powerful wind farm in the United Kingdom (County Kent) generates P = 300 MW.

The cost of wind energy decreases annually by 15% and even now can compete in the market; what really matters, it has perspectives for further cost decrease in contrast with the cost of electric energy generated by atomic power stations (which increases annually by 5%); moreover, at present the annual rate of growth of wind power exceeds 25% (Fig. 5.11).

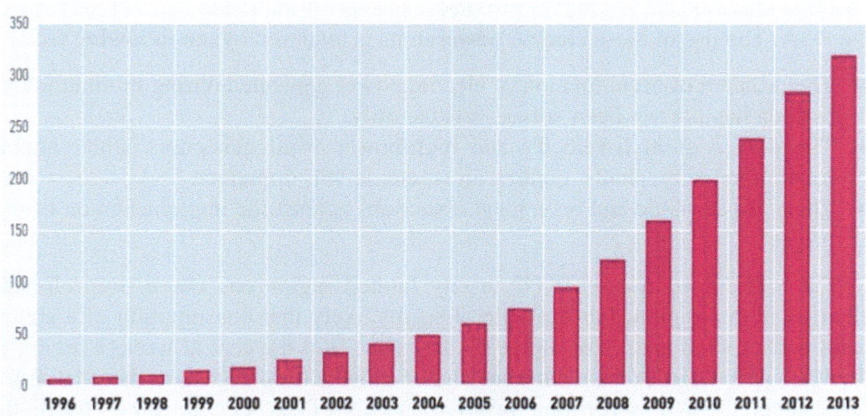

Fig. 5.11 Dynamics of installed capacity of WPP in the world, thousand MW

The process of wind power consumption intensifies in different countries. In 2014, 85 countries used wind energy on a commercial basis. To the end of 2015, in the windy industry were engaged more than one million people worldwide. In some countries, especially intensively developing wind power engineering, in particular, in 2015 in Denmark by means of wind turbines was produced 42% of all electricity; in 2014 in Portugal—27%; in Spain—20%; in Ireland—19%; in Germany—8%; in the EU—7.5%. Governments of China and city Tianjin (population of 14.5 million people) decided, that in 2020 more 20% of the electricity consumed by this city have to be produced with the help of WPP.

The average unit capacity of WPP operating all over the world makes ~ 140 kW. Until the middle of the 80s, WPP were based on wind-turbine units (WTU) rated at 100 kW. Since the middle of the 80s, WTU rated at 100–300 kW have been constructed, and since the late 80s, WTU rated at 600–700 kW have been built. At present, new models of WTU rated at 500–2500 kW are constructed based on the accumulated experience. The most common of modern commercial WTU with horizontal axis have a power of 2500 kW. A prototype of wind power generator rated at 6000 kW with a wind wheel about 130 m in diameter has been constructed. Almost all world stock of WTU consists of impeller units. The work is underway on WTU of other types and on impeller WTU with higher capacity. However, they are not widely used, and the perspectives for their application are not clear.

Thus, by the present time, the world wind power engineering has turned into the branch of electric power engineering making significant contribution to electric power generation in some countries.

One of the projects developed in the USA envisages the construction of 150,260-m towers with three-blade rotors turning round generators with a single capacity of ~ 1.5 MW. The density of location of these WTU will be one per square mile. As a result, their aggregate capacity will be about 225 GW, which is an essential share of the aggregate installed capacity of all electric power stations of the USA. The use of these electric power units is hindered by the following factors:

- The problem of accumulation of electric power generated during minimum load periods has not yet been solved satisfactorily.
- The design of high-velocity and high-power wind generators with variable rotation velocity, shafts, control units, etc. is still imperfect.
- There are aesthetic and ecological objections against the implementation of this project.

The widespread use of WTU in any limited region can cause deep climatic changes in this region. For example, it seems likely that consumption of a significant part of wind energy of a gale ($\sim 4 \times 10^{12}$ J) in the central western region of the USA will decrease the recurrence and wind velocities of gales at these latitudes. Of course, there is something positive in this effect. On the other hand, the remaining existing features of the climate in central western regions can also depend on gales. A decrease in the gale wind velocities can change the precipitation regime to such extend that some territories in eastern regions of the USA will become unsuitable for agriculture, and irrigation will be necessary for the remaining regions, as in the Far West of the country. The interaction of different atmospheric phenomena is a very complex process that has not yet completely understood. Any large-scale changes of natural phenomena on the Earth must be made carefully with allowance of their possible environmental effect.

Summarizing, we conclude that the wind energy is useful as an additional source of electric power, but in the nearest future, it will find only limited application. Scientific and technical progress, especially in the field of electric energy accumulation, can change the situation; however, in this case it will be necessary to evaluate the influence of large-scale wind energy consumption on the climate.

5.5   Solar Power Engineering

The Sun is the source of life on our planet and the source of all types of energy produced on it. The person has long paid and pays repeatedly attention to the utilization of solar energy to produce commercially electric energy, hot water, and vapor.

5.5.1   Electric Energy Production

At present, solar power plants (SPP) of two different types, distinguished by methods (thermodynamic and photoelectric) of solar energy conversion into electric

Fig. 5.12 Solar power engineering: *STPP* solar thermodynamic power plants, *SPVPP* solar photovoltaic power plants, *BECC* bio- and electrochemical converters

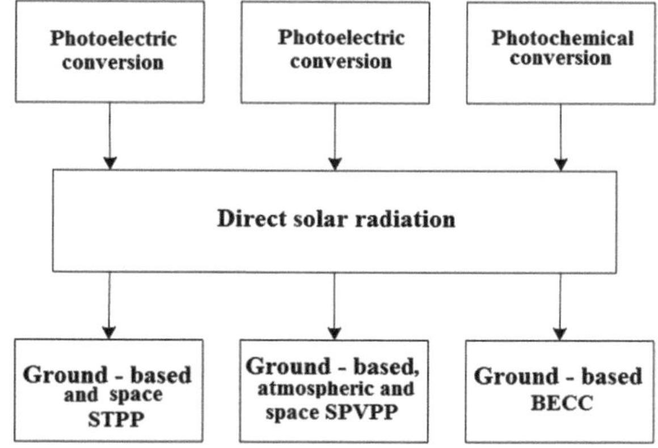

energy, are built and are in service (Fig. 5.12). (The third method—photochemical conversion was considered above.)

In the first case, solar radiation is converted into heat with rather high potential, then into mechanical energy (in a turbine or other heat machine), and finally into electric energy (in a generator). The photoelectric method is based on direct conversion of photon energy into energy of current carriers in irradiated semiconducting photoelectric pickoffs called the photoelectric effect.

The photoelectric effect was discovered by Hertz in 1887 and was studied in detail by A.G. Stoletov in 1888. Practical application of the photoelectric effect for producing the electric power has become possible only recently due to progress in physics of semiconductors.

When semiconductors with the electron (n-type) and hole (p-type) conductivities are in contact, a contact potential difference arises at the interface owing to electron diffusion. When the semiconductor with the p-type conductivity is illuminated, electrons in it absorb light quanta and go over to the semiconductor with the n-type conductivity. In this case, an electric current will run through a closed circuit.

Now silicon photocells illuminated by both direct solar rays and scattered light have the best characteristics. The efficiency of silicon photocells increases as temperature decreases, that is, they can equally successfully operate in both winter and summer. In winter, the decrease of the light flux is compensated by the increase in their efficiency at the expense of intensification of the photoelectric effect.

Because of complicated production process of semiconductors and their high cost, silicon photocells are used only for unique equipment, for example, of the Earth's satellites and space stations.

Despite the radical difference in the methods of energy conversion, solar thermal (STPP) and solar photovoltaic power plants (SPVPP) have a number of properties and limitations in common caused by the nature of the employed power source. Solar radiation as a power source, in addition to such positive properties as practically unlimited resources, complete ecological cleanness, and presence everywhere has also negative properties including *low density (specific power) of solar radiation* (no more than 1 kW/m^2 on the Earth's surface) *and objective (daily and seasonal) and random (caused by weather conditions) time variations.*

Already implemented and only developing ideas can simultaneously reduce the impact of shortcomings in the development of solar energetic. There are several ways to get the required amount of electricity and heat in conditions of a low surface density of the solar energy:

- predominant use of small power plants, allowing to create architectural and engineering compositions that organically combine the natural landscapes and habitats with the power plants. These SPP can form a spatial architectural compositions which are elements of the facades and roofs of buildings of almost any designation;
- installation panels of large SPP at a height of 2–3 m above the ground, allowing the land under them to be used for agricultural purposes (e.g. for grazing), or placing panels on land not suitable for other applications (e.g., deserts),
- construction of floating marine-ocean based SPP,
- creation of a balloon SPP which does not require a large area of land; placing a balloon above the clouds at the same time helps to solve another problem— dependence of the SPP power on the weather.

The irregular arrival of solar radiation to the Earth's surface leads to uncontrollable energy generation with the help of the SPP. Only the probable output power of the SPP in the given time of light day can be estimated based on the long-term meteorological observations. Other plants of interconnected power systems can compensate for this disadvantage, because a relative contribution of SPP to the output capacity of these systems is sufficiently low. In future, if the share of SPP in the electric power generation significantly increases, this disadvantage can become important. Scientists and engineers are trying to overcome these negative properties of solar radiation in four main directions: (1) the construction of giant solar panels in areas of the earth surface is not suitable for other applications (Fig. 5.13) or on the surface of oceans, seas, and big lakes, (2) placing solar panels on the Earth orbit or on the surface of the Moon, (3) more efficient solar radiation capture and converting it into electrical or thermal energy, (4) accumulation of energy during periods of excess electricity production or in the form of primary (electricity) or converted into other forms of energy: chemical (hydrogen produced by electrolysis of water), potential energy of water (pumped storage hydroelectric power plants). The real way to overcome the temporary instability of the solar energy input to Earth is the construction of the so-called "solar-fuel power plants". The essence of this concept is a combination of a solar thermal power plant with a maneuverable fuel unit compensating for the disadvantage of thermal energy supplied to a steam generator when the solar energy is lacking or insufficient. Such combined solar-fuel plants have sufficiently stable characteristics and can balance electric power.

The perspective direction is space-based solar power engineering (up to 10 TW of aggregate capacity) with energy transfer to the Earth surface. The available

Fig. 5.13 Ivanpah Solar Electric Generating System (Desert Mochava, USA). Characteristics: power—392 MW, footprint on the Earth—13 km^2, mirrors quantity—350 thousand, boilers at height of 140 m, cost—$2.2 billion

scientific and technological potential of astronautics creates premises for solving the problem of utilization of solar energy by several methods. They are:

- power supply of the Earth's regions with the help of orbital power plants converting the solar energy into microwave or laser radiation and transmitting it on the Earth. In future with increase in the number and capacity of these power plants, it will be economically expedient to construct them from lunar materials and to locate them on the Moon (see Fig. 5.14);
- lighting of near-polar regions of the Earth with the help of orbital solar reflectors made from thin film mirrors;
- increase of biomass production on the Earth by increasing the duration of light time of day;
- increase of the electric power generated by solar ground-based stations at the expense of additional illumination;
- heat supply to ground-based energy-technology complexes with the help of laser radiation generated on the Earth's orbits by conversion of solar radiation.

The specific mass of full-scale SPVPP rated at a few GW having classical configuration with silicon photoelectric converters and a microwave system based on electronic vacuum devices (magnetrons and amplitrons) with unit power in the range 100–1000 kW and moving in a geostationary orbit (at an altitude of 36,000 km) will be about 10 kg/W. It is expected that the efficiency of the entire process will be sufficiently high.

Space-based solar plants can be rated at 3–20 GW and even more. The size of the solar cell of the plant with an output power of 5 GW can be evaluated

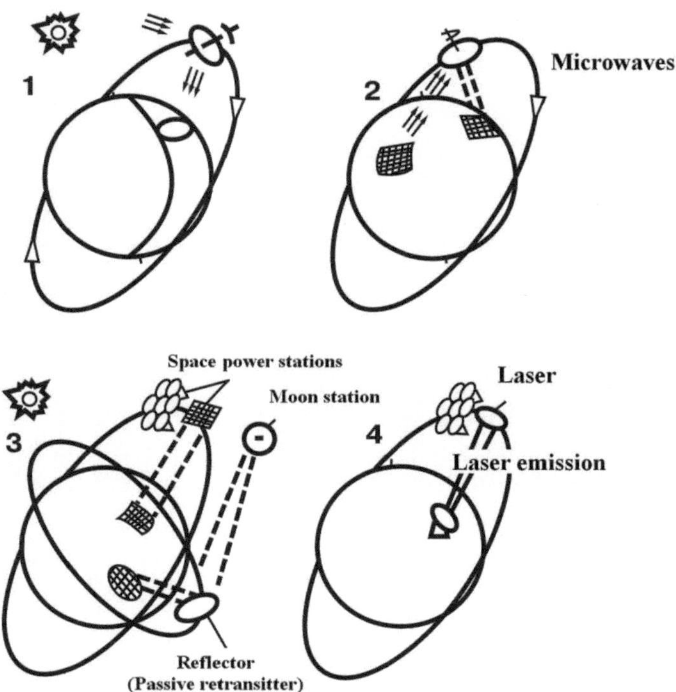

Fig. 5.14 Schemes of electric power supply of the inhabitants of the Earth with using of the Moon

proceeding from an efficiency of 15%. The surface area of the solar cell of this plant will be ~ 20 km^2. In this case, the transmitting antenna should have a diameter of 1 km, and the receiving antenna should have a diameter of 7–10 km. The energy density of the beam of ultra-short waves propagating from the space-based station to the Earth will be only 1/5 of the solar energy density; therefore, it will represent no hazard to flying vehicles or birds. The problem of radio interference will not be serious.

At present, there are no power limitations for SPP. In principle, SPP rated at any preset capacity, depending on the receiving antenna aperture, can be constructed. Therefore, such characteristics as, for example, the output electric power, is convenient to express per 1 m^2 of the receiving antenna aperture.

It is expected that *the efficiency of the conversion of solar radiation into electric power* will be sufficiently increased. Now the efficiency of solar energy conversion by single crystal cells is 11–12%; a 16% maximum efficiency was reported in the literature. It is assumed that 20% efficiency can be realized by improving semiconductor cells. Gallium arsenide single crystals provide an efficiency of 14%, and individual cells have an efficiency of 18%.

However, now in the laboratory is reached the solar energy conversion efficiency equal to 30–35%, and the possibility of its increasing to 40–45% in the foreseeable future is considered.

The share of solar modules on the base of crystalline silicon in annual photovoltaic panel's production is 93%. Most of the rest of the production—a thin-film amorphous silicon solar cells, including complex cascade; about 1% of the total manufacturing—the thin film solar module with solar cells that do not contain silicon such as poly-crystalline thin film elements based on CdTe and $CuInSe_2$ (CIS).

Improving solar cells (photoelectric converters) is conducted in several directions:

(a) Improving the efficiency of solar cells with the use of expensive single-crystal substrates and composite allows one to extend the part of the spectrum of sunlight converted into electricity. Solar cells are elaborated that convert into electricity not only light, but also heat. Radical increase of the efficiency solar of cells (40%) can be achieved by using crystalline silicon instead of GaAs.
(b) Reduction of the price of production technology of traditional materials, while maintaining the achieved level of efficiency. Despite the fact that silicon—the second most common element on Earth, the production of solar cells from it is expensive due to the complexity of the process of cleaning it from impurities. The cost of solar modules has decreased by 10 times over the past 30 years.
(c) Development of flexible (including optically transparent) solar cells, which makes their use easier and more versatile (for example, setting them in windows of houses).
(d) The use of redox reactions in the electrolyte, similar to the photosynthesis of plants; real efficiency of such conversion is 10%, and theoretical efficiency is 33%. Converters can be produced from cheap materials with the help of simple technologies.

A number of other criteria determines competitiveness of the SPP:

- the overall efficiency of the SPP should not be lower than 20%;
- SPP should generate energy as many hours per day as possible and throughout the year;
- SPP service life must be at least 50 years;
- the cost per installed kilowatt power should not exceed $1.000;
- the volume of production of the semiconductor material for solar cells should be more than 1 million tons per year at a price no higher than $12/kg;
- materials and components of the SES technology should be ecologically clean and safe.

Systems based on solar cells are essentially DC-sources. Up to 10 kV, AC power is obtained by means of an inverter connected, in general case, to a single-phase AC-grid. In the absence of a grid, and with low-quality requirements on the output current, we may connect the output of solar cell directly to an inverter, which supplies the consumer. In the presence of a grid, an additional DC-converter is

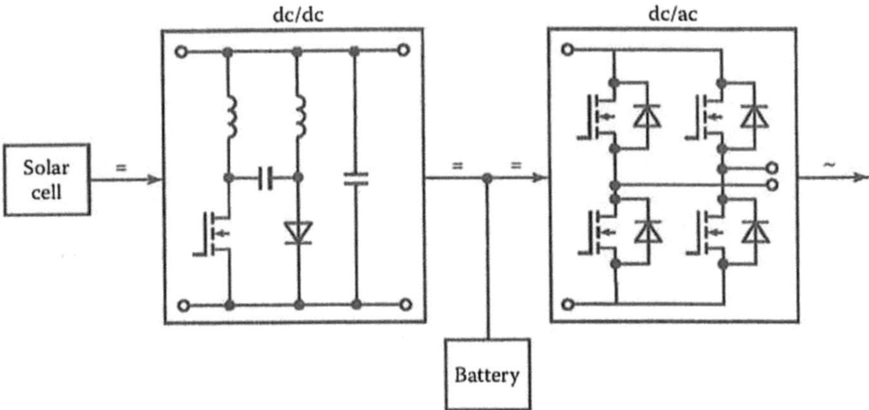

Fig. 5.15 Power circuit block diagram for a system with a solar cell

required (Fig. 5.15), because of the extreme instability of the output voltage from the solar cell. In addition, most systems include battery storage, to ensure stable voltage at the DC-bus, Fig. 5.15 [3].

The conversion efficiency of solar energy into electrical energy by a thermo-dynamic cycle depends on efficiency of transformation in all stages: the concentration of sunlight by using a system of mirrors, converting sunlight into thermal energy and thermal energy into the mechanical one and, finally, the mechanical energy into the electrical energy.

According to forecasts of the International Energy Agency (IEA), by means of solar energy about 5 thousand TWh of electricity will be produced to 2050, which is approximately 12% of the needs of all humanity. As the key to the implementation of these giant plans can be considered such a fact, that many oil and gas companies are considering solar power engineering as one of the promising trends in the diversification of their business. (For example, oil companies Shell, BP and others buy plants for the production of solar cells and execute in developing countries solar energy programs.)

Even today in countries with a predominance of sunny days per year, on the share of solar energy falls a significant part of electricity production. For example, in the Chilean capital Santiago, 60% of the electricity consumed by the underground is produced with the help of the SPP (and yet 16%—with the help of the WPP). The SPP with power of 100 MW is now constructed in Chile.

5.5.2 Thermal Energy Production

The solar energy can be used directly for heating water, buildings, and air conditioning. The advantage of solar energy consumption for these purposes is its absolute ecological cleanness. Getting thermal energy from the Sun by direct heating of the coolant (usually water) in solar systems is the most effective and economical way of solar energy conversion. The total cost for the construction of SPP, related to the collector area of 1 m^2), today accounts for about 400 euros; investments in the 1 kW are 230 euros. (If take thermal productivity of 1 m^2 solar panel equal to 1.75 kW (the average for the day.) This is 22 times less than the photoelectric conversion of the solar energy.

The most widely used four types of water heating systems by solar radiation are:

1. Single-seasonal solar installation of hot water supply (HWS). Installation with acceptable dimensions can provide heating of 100 m^3 water to a temperature of 55–60 °C, which is quite sufficient for domestic needs. They are used in seasonal objects: in cottages, summer recreation centers, sanatoriums, camping, etc. The service life is 15–20 years, and the payback period is 3.5–5 years.

2. Dual all-weather solar HWS installation. The first circuit of this installation is almost similar to a single-seasonal system, but is filled with liquid not freezing to—40 °C and is equipped with the necessary shut-off and regulating valves, circulation pump (10), and automatic control unit (4), Fig. 5.16.

   To heat water up to the technological standards in the cold season, there is a second circuit with an external heater (12) working on electricity or any form of traditional fuels. Here, the first solar circuit is an additional source of heat, which saves fuel or electricity.

3. Dual-circuit all-weather combined solar installation of HWS and heating is the second type of installation, retrofitting the boiler system, heating panels, expansion vessel, and circulation pump. Installation does not provide heating completely by solar energy, but solar circuit saves traditional energy resources.

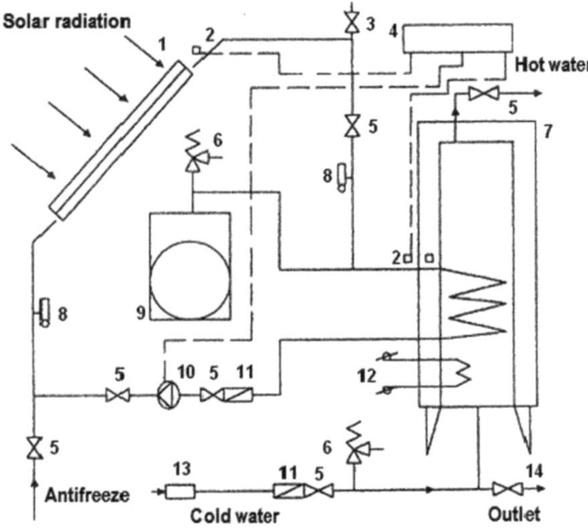

Fig. 5.16 Driving dual-HWS solar installation: *1* solar collector, *2* temperature sensors, *3* air valve, *4* automatic control unit, *5* valves, *6* safety valves, *7* water heater, *8* thermometers, *9* and *15* expansion vessels, *10* circulation pump, *11* non-return valve, *12* peak electric heating element, *13* water inlet, *14* exhaust valve

4. Single-contour all-seasonal solar power installation for HWS is an electric water
   heater equipped with a special solar power panel with cavitary absorbent for
   trapping solar energy and transmitting it to the heating water. All elements of the
   system including the tank—water accumulator—are located in a single housing.
   Such devices can be produced with capacity up to 120 L, which allows their use
   in apartment buildings. They can be built into the window openings, installed on
   balconies, loggias. Device with capacity up to 30 L can be produced in a por-
   table version, which infinitely expands the scope of their application, up to field
   conditions.

The effectiveness of solar thermal energetics continues to increase due to
improved performance of all elements of the system; in the best models of installations, the efficiency already reaches 80–85%.

Possibility to place solar cells and solar collectors on the roofs of buildings is
very handy in a densely populated country like Japan. When you have reached the
appropriate efficiency, it is necessary to place on the roof of the cottage only 6–
9 m^2 of solar collectors to meet needs of the average family of four in hot water and
heat.

In the USA, Germany, and Japan the number of homeowners who have solar hot
water and heating system has already reached several million people. In Spain, a
law according to which from January 1, 2005 each newly built house has to be
equipped with solar collectors was enacted. In several other countries: in Israel, the
UK, Switzerland, France, China, Greece, Portugal, Australia, and Cyprus a significant part of the needs of the residents of private homes in hot water and heating
is met by solar collectors. The total area of solar water heating systems in the world
exceeds 150 million m^2, most of which are built in China (59%) and Europe (15%).

Saving fuel consumed in huge volumes for heating in several northern hemisphere countries (Russia, the Scandinavian countries, Canada, etc.) can be provided
with large-scale use of passive solar heating systems in the construction of buildings. They provide a natural circulation of hot air and the required parameters of air
in the room without additional energy due to the following methods:

- use of helio-aktive external walling;
- use of thermal insulation;
- choice of materials for building construction with appropriate thermos-
  cold-accumulating properties;
- use in a heating system more thermo-cold-accumulating materials;
- equipping the premises with special elements (curtains, blinds, valves) allowing
  flexibility to adjust the air temperature in the room.

Considerable attention is given to the perspective of solar energy consumption in
the intermediate process of fuel production. Thus, problems of constructing large
solar stations the energy of which can be used for synthesis of hydrocarbon-based
fuel, for example, from methanol, limestone, and water were discussed at different
scientific conferences. This liquid fuel will allow one to avoid the problem of

storage and transmission of electric power at long distances. The liquid fuel can then be distributed and used as a conventional one.

To overcome the energy crisis, considerable attention is also given to an analysis of the role of solar energy in future power engineering. Already today solar power engineering is developed at an accelerated pace. Since the early 90s, the annual increase of the share of solar power engineering is 16%, while the annual oil consumption in the world increases only by 15%.

Solar power-generating units are most widespread in some EU countries (Greece, Portugal, Spain, and France), in the USA, Japan, Israel, Australia, and China. Each of these countries has from a few to tens of millions of square meters of solar panels. Nowadays, 1 m^2 of a solar collector generates: electrical energy— 1070 to 1426 kWh a year, heat—130 to 175 L of water to a temperature of 50 °C.

It saves annually:

- 1070–1426 kWh,
- 110–145 m^3 of natural gas,
- 0.18–0.24 ton of coal,
- 0.95–1.26 ton of firewood.

Solar collectors with an area of 2–6 million m^2 provide generation of 3.2– 8.6 billion kWh of energy and save 0.42–1.14 million ton of standard fuel annually.

Undoubtedly, interest in utilization of solar energy—the main energy source— will further increase, and a realm of its application will expand, Fig. 5.17.

To increase the solar energy contribution into the saving of non-renewable energy resources and to reduce the negative impact of power engineering on the environment, scientists and engineers are considering others approaches not mentioned above. At present, many of them are perceived as exotic.

Among implemented in Russia projects belonging to solar energy, the most appropriate is the construction of SPS 10 MW in the Mountain Altai. Despite the low power, it is of great importance for the saving of this unique nature reserve.

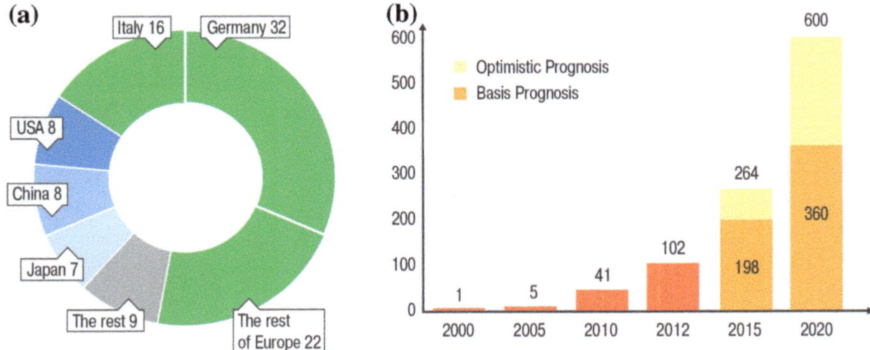

Fig. 5.17 Modern state (in 2012) and the forecast for the development of solar power engineering: **a** installed power of SPP in 2012, GW; **b** dynamics of installed power of SPP, GW

Construction of TPS or hydroelectric power stations on small rivers could cause irreparable damage to nature.

5.6   Tidal and Wave Power Plants

5.6.1   Tidal Power Plants

The energy of marine tides sometimes called the "lunar energy" has been known since antiquity. Already in past historical epochs, this energy was used to actuate various mechanisms, in particular, mills. In Germany, fields were irrigated using the tidal wave energy, and in Canada it was used to saw woods. A tidal lifting machine supplied London with water in England in the 19th century.

The potential of electric power generation from *marine tidal currents* is enormous.

The total tidal energy of the Earth surpassed full of river hydropower and estimated at 3 billion kW. The potential of tidal energy is estimated at 15% of the current electricity consumption.

Considering tidal currents, osmotic and thermal energy developments are recognized as a resource to be exploited for the sustainable generation of electrical power. The high load factors resulting from the fluid properties and the predictable resource characteristics make marine currents particularly attractive for power generation and advantageous when compared to other renewable energies. Moreover, international treaties related to climate control have triggered resurgence in the development of renewable ocean energy technology. Therefore, several demonstration projects in tidal power are scheduled to capture the tidal generated coastal currents [4].

There are a huge number of witty projects of tidal equipment. Only in France more than 200 inventions had been patented by 1918. In the early 20th century, some attempts were undertaken to construct large tidal power plants (TdPP). The construction of the Quoddy TPS rated at 200 thousand kW was started in the USA in 1935. Soon the construction, on which 7 million dollars had already been spent, was stopped, because it was found that the cost of electric power would be too high (it would exceed by 33% the cost of electric power generated by heat power stations). According to the project developed in the USSR in 1940, the Kislogubsk tidal power station would produce energy whose cost would be twice as great as that of river power stations.

The advantage of TPS over river power stations is that their operation is governed by space phenomena and is independent, unlike the river power stations, on numerous random weather conditions.

However, TdPP have two significant disadvantages, namely, irregular operation and large volume of investments.

The irregularity of tidal energy within the lunar day and lunar month that differ from solar ones does not allow TPS to be used systematically during periods of maximum loading in power systems. The irregularity of TdPP operation can be compensated by their combination with pumped storage power plants (PSPP). When there is an excess output power of tidal power stations, PSPP operate in the pump mode consuming this excess power and pumping water to the upper reservoir. During periods of minimum output power of tidal power stations, PSPP operate in the generation mode supplying the electric power to the system. This project is technically acceptable but expensive, because large installed capability of electric machines is required.

The tidal power stations can also be combined with river HPP having a storage reservoir. When these stations operate together, the HPP increase their output power when the output power of tidal power stations decreases or when they are stopped; when the tidal power stations generate sufficiently high power, PSPP pump water to the storage reservoir. Thus, both daily and seasonal irregularity of operation of the tidal power stations can be compensated.

Work cycle of the simplest (classical) TdPP includes filling the pool with water during high tide through holes in the dam, inclusion the turbine and connected to it the electric generator at low tide when the water pressure becomes sufficient.

Today is considered cost-effective construction of TdPP in areas with tidal fluctuations in the water level of at least 4 m. Efficiency of TdPP is possible to increase the use of turbines, working during both high tide and low tide. Such TdPP are able to generate electricity continuously for 4–5 h with intervals of 1–2 h four times per day. To increase the operating time of the turbine, there are more advanced schemes—with two, three and more pools, but the cost of these projects is very high.

The TdPP operate under conditions of fast turbine head change; therefore, their turbines must have high efficiency at variable heads. A rather good compact horizontal double-flow turbine is now in the market. An electric generator and some parts of the turbine are enclosed in a watertight capsule, and the hydraulic unit is immersed in water. Rotating vanes of the turbine provide high efficiency at different heads starting from 0.5 m.

The hydraulic unit can operate both in generation and pump modes. When the generator is switched off, the hydraulic unit can pump water directly from the sea to the reservoir and back; in the pump mode, it can pump water from the sea to the reservoir and by that to increase the head.

The nature of Russia allows tidal power stations with an aggregate installed capability of about 150 thousand MW to be built. Long-term scientific investigations and projects have led to a conclusion that of interest is the construction of a few tidal power stations, rated from several hundred to several thousand MW. Three powerful tidal stations operate: (a) in France (the Rance tidal power station rated at 240 MW in France having 24 units; put into operation in 1967), (b) in China (the Tsansyan tidal power station rated at 3.2 MW; its six units were launched in 1980–1985), and (c) in Canada (the Annapolis tidal power station rated at 19.6 MW, having 1 unit built in 1984). In addition, tens of tidal power micro- and mini-stations being integral parts of complexes used for irrigation, drainage, navigation, etc. were built in China.

A large number of tidal power station units are severe obstacle for their construction, because to build such a number of units, it is necessary to involve all power industry of the country.

5.6.2 Wave Power Plants

A promising source of renewable energy sources are waves of the oceans, seas and large lakes that can develop the greatest specific power, among others renewable energy sources. The waves of all oceans, sees, and big lakes in the future are able to provide up to 2 TW of electric power, enough to meet the needs of all mankind. With the present level of technology, wave power plants (WavePP) with a total capacity of up to 10 billion kWh can be created.

The trend in the development of WavePP as power plants using other energy renewable resources—a complex of single intermediate power modules (approximately 1 MW). A WavePP module has a size of about 50 m along the wave front. Complexes with hundreds of meters in size and the total capacity of tens of megawatts are often constructed of these modules. The low-power modules (tens of kilowatts) with a sufficiently broad consumer niche had been developed. As in the case of tidal energy, a number of technical solutions currently proposed and partially implemented can be grouped into two groups faced on prevalence of kinetic energy or energy rolling surface, which is converted into electrical energy. By design, WavePP can be divided (rather conventionally) on the float-WavePP and

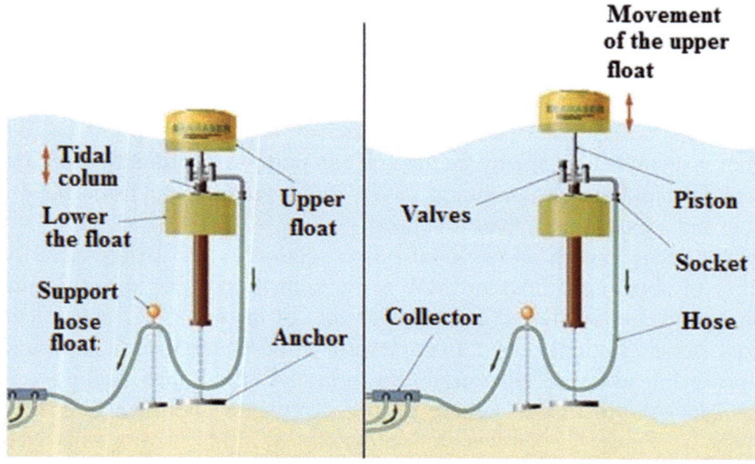

Fig. 5.18 Principle of operation the float-WavePP

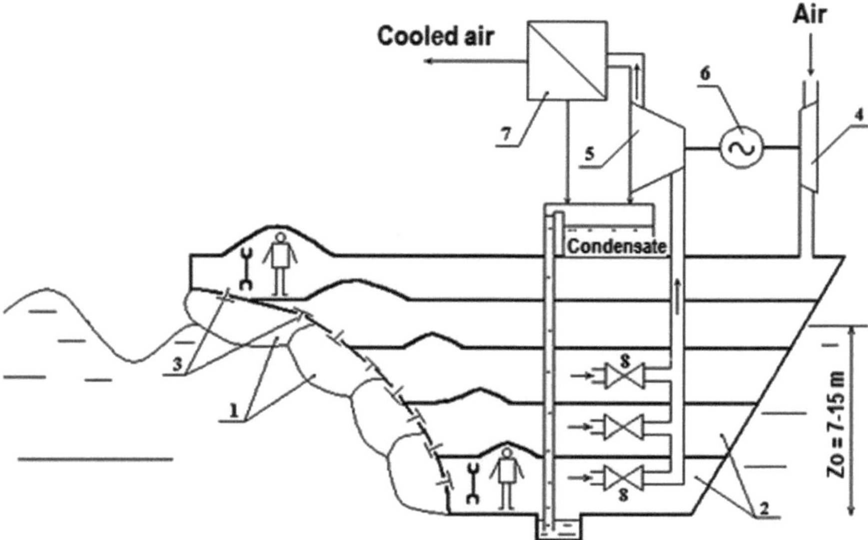

Fig. 5.19 Principle of operation ща the chamber-WavePP: *1* wave-capture camera; *2* receivers; *3* check valves; *4* turbine; *5* generator; *6* moisture separator; *7* compressor; *8* shut-off valves

chamber-WavePP. The principles of their work are shown in Figs. 5.18 and 5.19, respectively.

There are two main variants of the chamber. One of them is WavePP in which the kinetic energy of waves used. In this variant, a reception pipe of a very large diameter (in the general case—the chamber of large cross section of any shape) is established in the way of the waves. Incoming waves rotates a turbine which drives a generator.

Another variant of the WavePP is because tidal waves passing through special chamber displace the air contained in this chamber. Compressed air under pressure passes through the turbine rotating it and electric generator. As a result, electricity is produced.

Figure 5.20 shows the photograph of the chamber-WavePP.

Tests of a WavePP prototype have demonstrated that such system can operate not only in the coast where waves are permanent, but even in large lakes. The inventors believe that it is expedient to connect such units in series to form a battery thereby creating a reliable and cheap source of high electric power. Of course, these reasoning must be confirmed under operating conditions.

Fig. 5.20 Photograph of wave power plant "Oceanlinx" (Australia)

5.7   Geothermal Power Plants

Globally, one of the largest sources of sustainable and environmental friendly heat that can be used for electric power generation is geothermal resource. Even so, owing to the lower steam temperatures (as compared to fired systems), plant efficiencies have been limited (to around 15%), which has correspondingly limited the economic lower range of geothermal temperatures that can be used. However, the current and projected rising cost of power generation is constantly pushing down that lower limit boundary, such that new low-grade geothermal sources and waste heat from existing fired generation or industrial process can be considered economical for power generation. As the temperature differentials reduce, different technologies must be considered when trying to meet commercial requirements.

According to the methods of the Earth heat extraction, of the following classification of geothermal systems is used, Fig. 5.21.

It is well known that the temperature increases by about 1 °C as the Earth's depth increases by 30–40 m. Therefore, water boils at depths of 3–4 km, and the Earth's temperature reaches 1000–1200 °C at depths of 10–15 km. In some regions of our planet, the temperature of hot natural springs is sufficiently high even near the Earth's surface. These regions are most favorable for the construction of geothermal power plants (GTPP). Thus, geothermal stations generate 40% of electric power in the New Zealand; 6% of electric power is generated by geothermal stations in Italy. These stations in other countries generate a significant part of electric power.

Italy was the first country that started industrial utilization of heat of the Earth's bowels. This was promoted by the lack of conventional energy resources there.

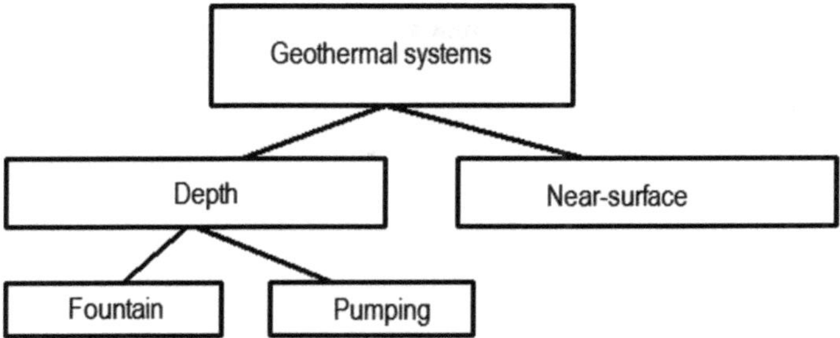

Fig. 5.21 Classification of geothermal systems

Fig. 5.22 Block diagram of a geothermal power plant for volcanic regions: *1* well (hole), *2* steam converter, *3* turbine, *4* condenser, *5* pump, *6* water heat exchanger

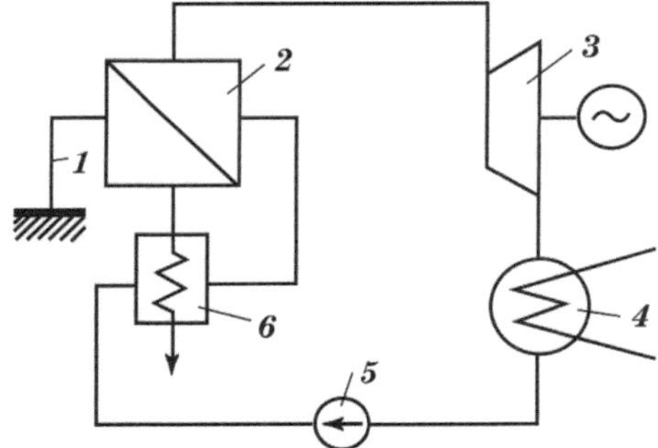

The scheme of underground heat utilization is simple. Hot underground water produces vapor utilized in geothermal power stations and other technical facilities. Vapor of the Earth's bowels, unlike steam produced by steam generators of TPP, contains impurities of different aggressive gases that destroy the equipment of stations. Therefore, vapor of bowels is either preliminary directed to heat exchangers to obtain pure vapor or special corrosion-resistant equipment is used. In the first method, about 25% of heat is lost. The second method is considered as most expedient now.

The base mode of operation is preferable for such GTPP, since wells allow no sharp changes in pressure and discharge.

The examined GTPP are geographically attached to natural springs of hot vapor and water; therefore, regions of their utilization are limited. GTPP on thermal water with temperature in the range 100–2000 °C can become more widespread; these GTPP should have two contours with a working agent of the secondary contour boiling at low temperatures.

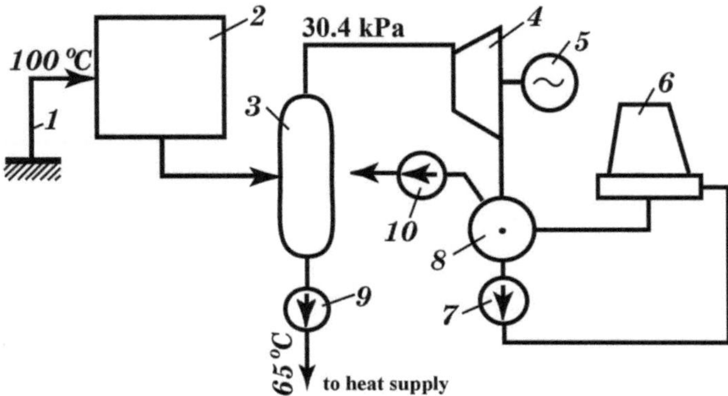

Fig. 5.23 Block diagram of a GTPP for volcanic regions: *1* well (hole), *2* tank-accumulator, *3* evaporator, *4* turbine, *5* generator, *6* water-droplet cooler, *7* pump, *8* condenser, and *9* and *10* pumps

The potential reserves of thermal waters with the above-indicated temperatures are concentrated in Northern Caucasus in water stratums at depths of 2.5–5 km and provide the basis for the construction of GTPP with an aggregate capacity of a few million kilowatts. However, whereas GTPP on natural springs of hot vapor and water are commercial facilities, GTPP on thermal water with the above-indicated potential require testing. Similar GTPP can also be combined with vapor hydrothermal GTPP to utilize the heat of separated water. This can increase electric power generation by $\sim 20\%$. The wells in fields of thermal water admit the discharge control; therefore, the output power of double-contour GTPP can be controlled without losses of the heat carrier.

Figure 5.22 shows the block diagram of a GTPP for volcanic regions.

Figure 5.23 shows the block diagram of a GTPP for volcanic regions with resources of thermal water having a temperature of 100 °C at depths accessible to modern drilling engineering, and Fig. 5.24—typical appearance of GTPP.

In the distant future, high-temperature (up to 1000 °C) mantle layers will produce vapor from water pumped to an artificially created volcanic mouth. Of course, the energy so produced will be pure and will not influence the biosphere (the huge mantle mass eliminates the effect of heat taken away on the mantle state).

Fig. 5.24 Typical external view of the GTPP

5.8   Other Renewable Energy Sources for Electricity Production

Among these energy sources are: (1) sea and ocean currents, (2) temperature difference between surface and deep sea and ocean waters in tropical latitudes, (3) air/sea temperature difference in the Arctic (Antarctic) region of the Earth, (4) distributed (low-potential) heat of sea and ocean waters and other large reservoirs, and (5) phenomenon referred to as *osmosis*.

5.8.1   Ocean and Sea Currents Energy

Global ocean currents, such as Gulf Stream, Curoshio Current, and Florida Stream transport water at rates of 83, 55, and 30 million m^3/s, respectively, with velocities of 2, 1.8, and 1 m/s and possess huge energy. From 1 m^2 of the cross sectional current area, a capacity of about 1 kW can be obtained. Currents in Gibraltar, La Manche, and Kuril Straits as well as tidal currents are also of interest for power engineering.

One of the methods of conversion of this energy into electric one is implementation of the idea that combines two technical solutions—a wind-wheel and a tidal power station unit. This allows hybrid HPP operating in free running water to be built whose units possess advantages of both wind-wheel and tidal power station:

(1) They have much smaller water propeller sizes and hence, reduced material and labor consumption. To receive a power of 1 MW, aqueous two-blade propeller diameter should be about 18 m, and the propeller—about 55 m. The aqueous two-bladed rotor with a diameter of 1 m at a flow rate of 2 m/s can provide power of 7 kW.
(2) Compared to TdPP and HPP, the unit of installed capacity for them is by about an order of magnitude cheaper (primarily due to the absence of a dam).
(3) They allow operational characteristics to be estimated much more correctly due to a better predictability of the sea current parameters in comparison with the wind parameters.
(4) They are invisible and do not change the landscape and the natural state of the coastal line. The hydro turbine (comprising a water propeller and an electric generator) is mounted on a completely submerged support and is lifted out of water only during repairing or servicing, Fig. 5.25.

The world's first commercial electric power station of this type (operating in free running water) with a capacity of 1.2 MW has been connected to the National Electric Grids of Northern Ireland. The dual rotor turbine is used. Rotors of the turbine are 16 m in diameter and rotated with an optimal velocity of 14 rpm. The rotor blades are equipped with a control system and can be rotate to change the angle of attack. Marine Current Turbine Ltd. that built this power station has already started a new project on building a power station at the coast of Northern Wales with a capacity of 10.5 MW.

In the USA, the program Coriolis had been developed which envisages installation of 242 turbines in the Florida Strait; every turbine will be equipped with two driving wheels 12 m in diameter rotating in opposite directions. The whole system with total length of 60 km will be oriented along the main current; its width with turbines installed in 22 rows 11 turbines in every row will be 30 km. The units will be submerged at a depth of 30 m not to interfere with navigation. The capacity of each turbine will be 400 kW, and the entire complex will generate about 100 MW.

Intensive works on the development of hydropower stations intended for operation in free running water are underway in France that has great possibilities for the development of renewable electric power of this type, since strong ocean currents with velocities up to 15 m/s, caused by tides, run near beaches of Normandy and Brittany. Building of three electric power stations is planned: two in Brittany with a capacity of 1 and 2 GW, respectively, and one on the seaside of the Cotentin Peninsula. The total output of these stations will be about 25 thousand GWh that will make 5% of the total installed electricity produced in France. The payback period for the electric power stations is 7 years.

The firm BioPower Systems (Australia) has developed a project of the oceanic BioWave underwater power plant inspired by nature. Its design mimics the swaying motion of kelp plants. The external view of these units resembles the seaweed with

Fig. 5.25 External view of units that harness the energy of sea (ocean) currents

large flexible leaves continuously swaying back and forth with the surging waves. The pilot system of the power station with a capacity of 250 kW has successfully passed tests at the coast of Tasmania. In the nearest future, such power stations will provide electricity to Australian State Victoria including its capital Melbourne.

5.8.2   Thermal Energy of Ocean and Sea Water

To use the temperature difference between surface and deep sea and ocean waters in tropical latitudes, warm ocean water (24–32 °C) is directed to a heat exchanger where liquid ammonium, Freon, or propane are converted into steam that rotates the turbine and then enters the subsequent heat exchanger for cooling and condensation by water with a temperature of 5–6 °C taken from depths of 200–500 m. The electric power generated by such power plants (called ocean thermal energy converter (OTEC) stations) can be transmitted to the coast or consumed in the place (for example, in oil and gas platforms to extract mineral raw materials from sea water, to desalinate it, and so on). The idea of generating electricity in this way was made public as early as 1881. However, the first power plant of this type was built under the direction of French engineer Georges Claude in 1930 in Cuba.

This unconventional renewable power source (URPS) possesses important advantages over many other sources: (a) the temperature difference of various layers of ocean water is a more stable energy source than the wind, the sun, sea waves, and small river discharges; (b) such factory-assembled power stations can be easily delivered to any region of the World Ocean; and (c) their energy potential is estimated to be about 10^{13} kWh/year.

Their wide application is hindered by two serious disadvantages: (a) geographic reference to the tropical latitudes (between 20 and 29 S), where water temperature at the ocean surface reaches, as a rule, 27–28 °C, and where such power stations are economically efficient and (b) large mass and overall dimensions. The first experiments on harnessing of NRES of this type within the framework of the joint project of the Japanese and Indian scientists gave positive results. As far back as in the 70s of the last century the developed countries started the design of pilot OTEC stations representing complicated large-size structures with installed capacity of several ten to several hundred megawatts.

The main units of the OTEC are an evaporator to produce steam of working fluid by means of heat exchange with seawater, a turbine for the drive of an electric generator, a condenser for spent steam, and pumps for pumping sea water and cold air. The large specific dimensions (per unit generated capacity) are caused by low efficiency of the thermodynamic cycle of the OTEC station (the Rankine cycle). For this reason, the OTEC station consumes huge amounts of warm and cold waters of the order of several thousand cubic meters per second. Thus, the floating OTEC with a capacity of 40 MW should have 70 thousand tons displacement, cold water pipeline diameter of 10 m, and a working surface of the heat exchanger of about 45 thousand m^2.

In the last few years, projects of an OTEC with relatively small capacity have been developed in which already commercially available main units (turbines, electric generators, evaporators, condensers, and pumps), placed in a 20 foot (6.1 m) container that can be delivered to any required place, can be used. The installation that falls within these clearance limits can provide a capacity of 300–350 kW.

Recent discovery of hydrothermal sources on the ocean floor (with temperatures up to +350°) gave rise to the idea of building underwater OTEC stations harnessing

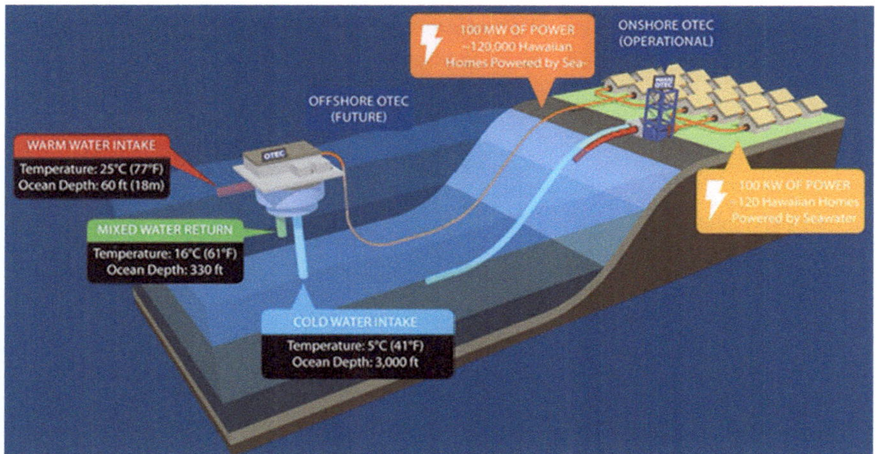

Fig. 5.26 Simplified diagram of OTEC constructed by Makai Ocean Engineering Company

the temperature difference between the sources and surrounding waters. A much greater temperature difference between hot water near geysers on the ocean floor and remaining water masses will provide much higher efficiency of the thermal station. However, extreme exploitation conditions of such OTEC stations can bring to nothing this advantage.

Recently, the US Company Makai Ocean Engineering built in Hawaii the world's largest facility of this kind, capable of generating electricity from temperature differences. It is planned that the plant will provide power about 120 kW nearby households. Its simplified diagram is showed in Fig. 5.26.

In the majority of regions of the Arctic Ocean the average long-term winter (November–March) air temperature does not exceed 26 °C. Warmer freshwater discharge of large rivers (such as Yenisei, Lena, and Ob') heats seawater under ice to +3 °C. Electric power stations harnessing this energy can operate by the conventional scheme based on the closed cycle with working liquid boiling at a low temperature.

Large positive experience has been accumulated on harnessing of low-potential heat by the heat pump scheme. The scheme is simple—seawater is heated to the temperature required for heating of hot water supply systems. For example, in Sweden Baltic seawater with temperature of +4 °C is heated to 75 °C using heat pumps and is then used for home heating. This appears more profitable than to use black oil or gas for heating.

5.8.3   Osmotic Energy

The idea of generating electric power taking advantage of the osmosis phenomenon (slow penetration (diffusion) of a solvent into a solution through a thin membrane

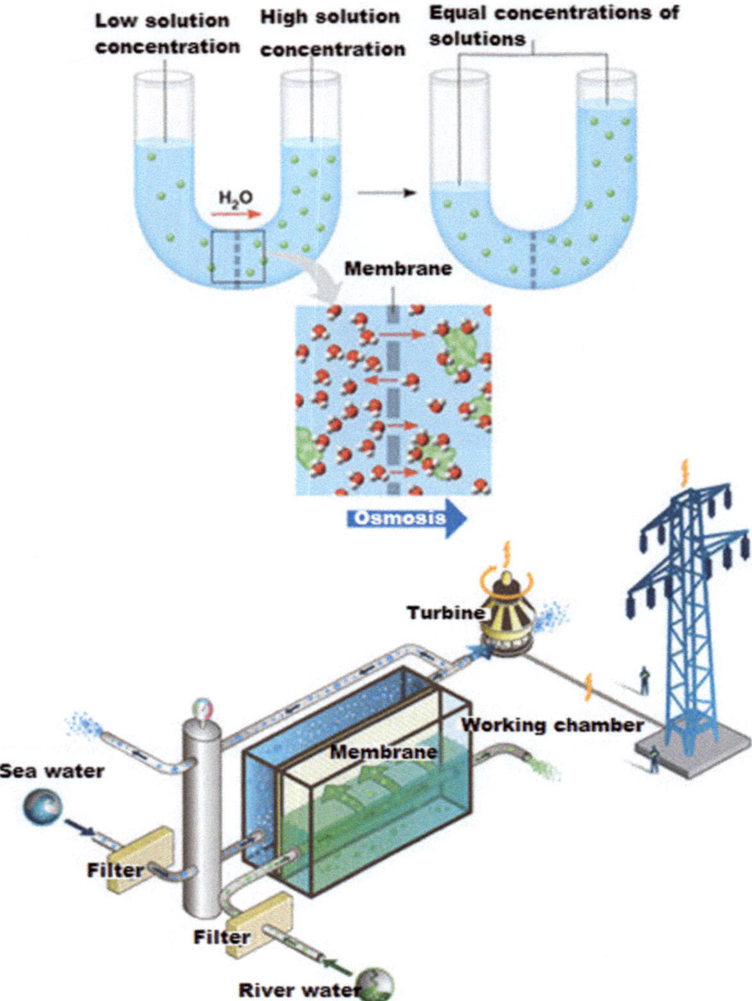

Fig. 5.27 Schematic representation of osmosis and structure of an osmotic power station

impenetrable for dissolved substances) well known in chemistry and biophysics was proposed 50 years ago. As applied to generation of the osmotic energy, the case in point is penetration of fresh river water from one half of a reservoir into another filled with sea water through a translucent membrane. As a result, in this half of the reservoir the pressure is created that can be equivalent to that of the water column 100–120 m high, Fig. 5.27.

The potential of the energy source of this type is 1600–1700 TWh/year. Power engineers are interested in this energy source due to its renewability, ecological purity, and sufficiently high stability of power characteristics (only insignificant seasonal variations are observed). For a long time it was impossible to implement commercially this idea, since the manufactured membranes were too thick and did not provide necessary osmosis rate and hence, efficiency of electric power generation. The world's first osmosis power plant (OPP), built by the Statkraft, a Norwegian electricity company, was put in operation with financial support of the state and some other companies only in November, 2009. The station with a capacity of only 5 kW is located 60 km from Oslo in the mouth of a river running into the North Sea and is in essence a pilot station intended to elaborate technical solutions to build a commercial power station within several years. The cost of the project including a study of the possibilities of osmosis application for electric power generation and building of a prototype was about 27 million dollars.

The main problem here remains the manufacture of very thin (<0.1 μm) commercial membranes which would provide power density at a level $\geq$ 5 W/m^2. The membranes for desalinization of seawater provide a power density of $\leq$ 1 W/m^2. The largest companies-manufacturers of the membranes (General Electric, Hydranautics, and Today Industries) can start manufacture of the membranes for the OPP no earlier than in 2015–2020.

In conclusion, to this chapter, it is necessary to underline that when solving the problem on expedient volumes of URPS harnessing, a multilateral approach is required in which the following factors should be taken into account:

(1) URPS technical and economic potentials in the given country and its regional distribution;
(2) Economic, ecological, and social efficiencies of power supply using NRES of various kinds which, in turn, are determined by such characteristics, as the cost of mineral fuel, average annual URPS power production, remoteness from the centralized power supply system, state of the road network, requirements of consumers to the quality of both electric and thermal energies, reliability and cost of NRES stations, etc. These characteristics should be evaluated in each concrete case.
(3) Technical feasibility of the project of URPS power supply.

The attitude to the NRES was radically revised in the West in the mid-70s primarily due to the adoption of a complex of regulatory legal acts that provided *the legal, economic, and organizational* bases for its subsequent development. The URPS power engineering was supported (and is continued to be supported) by both legislative and executive branches of governments. This is especially urgent in the stage of its promotion, formation, and adaptation in the power market.

Despite more than 30-year experience of state regulation and the progress made in this branch of economy, the advanced countries do not weaken and, on the contrary, strengthen their attention to NRES improving stimulation and support mechanisms. The effective financial and economic mechanisms of stimulation of NRES development in a number of countries are *green certificates*. They confirm the fact of generation of this or that amount of energy using definite NRES. The green certificates are produced by special bodies (one for each country), and their number corresponds to the amount of electric energy produced using URPS. Moreover, organizations that failed to comply with their obligations on URPS energy generation can report for their fulfillment by purchasing green certificates from organizations that produced excessive amount of URPS energy. The primary goal of the green certificates is to get additional source of financing for the development of ecologically clean power technologies.

The active state participation in the development of renewable power is also characteristic for many developing countries. Direct administrative regulation and economic stimuli are also used. In fact, in China, India, and some other countries funding is moved from the sphere of traditional power (mainly coal) to the NRES sphere. The development of renewable power, first of all independent, could become a solution to the problem of power supply in the poorest countries that do not have a modern power infrastructure—the developed network of the centralized power supply. However, they have no money to purchase sufficient quantity of the corresponding equipment.

The experience accumulated by the EU countries, the USA, Japan, etc. demonstrates that the problem of URPS harnessing can be solved only with active participation of private investors. To involve investors in business that does not bring fast profits, it is necessary to use the whole mechanism of stimulation created and approved by the countries—leaders within the last 30–35 years. Today the most effective mechanism of URPS project implementation is the private-state partnership. Business should solve the following problems:

- To estimate commercial attractiveness of projects,
- To provide marketable competitive production,
- To form the investment policy to organize mass production of the equipment and to create the RE infrastructure.

Table 5.16 gives the dynamics of global consumption of various renewable energy sources, including large hydropower ones. (According to the scenario developed by the European Council on the issue of renewable energy sources on the data basis of the International Institute for Applied Systems Analysis.)

Based on the data presented in Table 5.17, renewable energy future is quite optimistic.

Questions and Tasks to Chapter 5

1. What are major aspects of the fast increase of the energy consumption?
2. What do they implied by "nontraditional renewable energy resources"?
3. What is the main difference between nonrenewable and renewable energy sources?
4. Why is it necessary to expand the use of renewable energy? What is the current scale?
5. What are the main renewable energy sources?

Table 5.17 Increase in the world's renewable energy consumption volume, in million tons of oil equivalents and in percent

Volume of renewable energy consumption in million tons of oil equivalent and in percent	Years				
	2001	2010	2020	2030	2040
Primary energy	10038.3	10,549	11,425	12,352	13,310
Biomass	1080	1313	1791	2483	3271
Large hydropower	222.7	226	309	341	358
Small hydropower	9.5	19	49	106	189
Wind energy	4.7	44	266	542	688
Electricity from solar energy (solar panels)	0.2	2	24	221	784
Solar thermal energy	4.1	15	66	244	480
Electricity from thermal solar energy	0.1	0.4	3	16	68
Geothermal energy	43.2	86	186	333	493
Sea Energy	0.05	0.1	0.4	3	20
Renewable all	1364.5	1745.5	2694.4	4289	6351
The share of renewable sources (%)	13.6	16.6	23.6	34.7	47.7

6. Which countries are the leaders in the use of renewable sources?
7. What restrains the expansion of renewable energy?
8. Name the largest hydroelectric power plants in the world. Where are they located?
9. What limits the rate of development of large hydropower?
10. Advantages and disadvantages of micro- and mini-hydropower plants.
11. Storage power station (pumped storage plant): operating principles, advantages, disadvantages, and the current state.
12. What refers to bio-resources?
13. Name the method of using bio-resources in the energy sector.
14. What limits the use of peat in the energy sector?
15. Advantages and disadvantages of wind power plants.
16. Countries-leaders in the development of wind power plants.
17. Maximum power of existing wind turbines.
18. The main directions of solar energy application in the energy sector.
19. Advantages and disadvantages of solar energy.
20. Countries-leaders in the development of solar energy.
21. What are the principal reasons for the power supply of the Earth's regions by means of mirrors or orbital power stations?
22. The principle of operation of tidal power units.
23. Operating principles of wave power units.
24. Types of geothermal energy resources and use them.
25. Application of marine and ocean streams for energy generation.
26. Methods of utilization of thermal energy of sea and ocean water.

References

1. Hunt S, Shutteworth G. Competition and choice in electricity. Chichester, England: Wiley; 1996.
2. Rozanov Y, Ryvkin S, Chaplygin E, Voronin P. Power electronics basics (Operating principles, design, formulas, and applications). CRC Press; 2015.
3. Ben Elghali SE, Benbouzid MEH, Charpentier JF. Marine tidal current electric power generation technology: state of the art and current status. In: IEEE international conference on electric machines and drives 2007, IEMDC'07. 2007. vol. 2, p. 1407–12.
4. Short TA. Electric power distribution, equipment and systems. New York: CRC, Taylor and Francis Group; 2006.

Chapter 6
Energy Transmission and Distribution

6.1 Main Stages in the Development of Power Transmission Systems in the XXth and in the First Part of the XXIst Century

The initial stages in the development of power transmission systems were discussed in Chap. 2. Here we briefly describe the steps of further development of these systems in the XXth century and in the first decade of the XXIst century and organizational problems appearing in this process.

All countries now have a power system which transports electrical energy from generators to consumers. In some countries several separate systems can exist, but the growth in size of power plants and higher voltage equipment was accompanied by interconnections of the generating facilities. These interconnections decreased the probability of service interruptions, made utilization of the most economical units possible, and decreased the total reserve capacity required to meet equipment-forced outages. This integrated system (often known as the "**grid**") has become dominant in most areas and is usually considered as a major factor in the well-being and level of economic activity in a country. All systems are based on alternating current, usually at a frequency of either 50 or 60 Hz. (50 Hz is used in Europe, Asia (partly), India, Africa and Australia, and 60 Hz is used in North and South America and parts of Japan.)

This was accompanied by use of sophisticated analysis tools such as the network analyzer. Central control of the interconnected systems was introduced for reasons of economy and safety. The advent of the load dispatcher heralded the dawn of power systems engineering, an exciting area that strives to provide the best system to meet the load requirements reliably, safely, and economically, utilizing state-of-the-art computer facilities.

Already since 1901–1905 three-phase electric power stations, which at the beginning were predominantly stations of plant-factory type, have been built.

© Springer International Publishing AG 2018
V.Y. Ushakov, *Electrical Power Engineering*, Green Energy and Technology,
https://doi.org/10.1007/978-3-319-62301-6_6

Three-phase engineering allows large electric power stations to be built in places of fuel extraction, at waterfalls, or on a suitable river and to deliver generated electric power through transmission lines to industrial regions and cities. Such electric power stations were called regional. They provided the basis for the development of electric power engineering in the 20th century.

The first regional electric power station was the Niagara hydroelectric power station. Regional electric power stations have been built intensively from the beginning of the 20th century. This was promoted by an increase in the current consumption connected with increasing industrial applications of the electric driver, the development of the electrical transport, and electrical illumination of cities.

Electric power stations became large industrial enterprises that produced electric power; different stations were integrated, and first electric power systems were formed. By the electric power system, we mean a set of electric power stations, transmission lines, substations, and central heating systems having the common operation mode and uninterrupted process of production and distribution of electrical and heat energy.

It was elucidated that joint operation decreased the power reserve at each station, the equipment could be repaired without switching off the main consumers, and conditions arose for equalization of the production load of the main stations to increase the efficiency of utilization of power resources.

The subsequent consolidation of electric companies enabled economies of scale such as capital cost of generating facilities, the introduction of equipment standardization, and utilization of the load diversity between areas.

Extra high voltage (EHV) has become dominant in electric power transmission over great distances, Fig. 6.1.

The major ac transmission systems in chronological order of their installations are shown in Fig. 6.2.

The power transfer capability is approximately $P = V^2/Z_1$, which for an overhead ac system leads to the following results:

Fig. 6.1 Dependence of the power transmitted through three-phase overhead transmission line (OTL) from the operating voltage

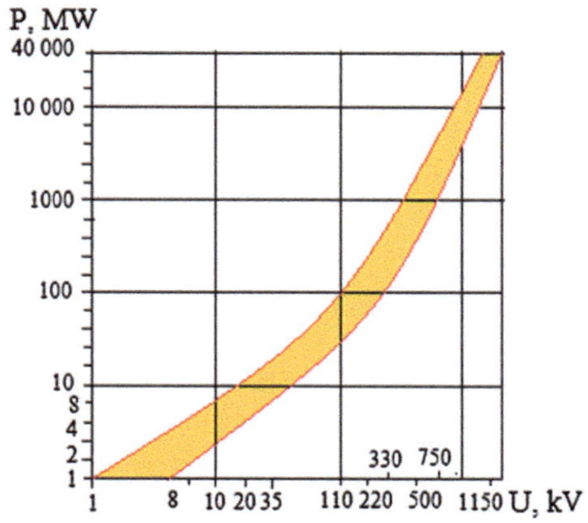

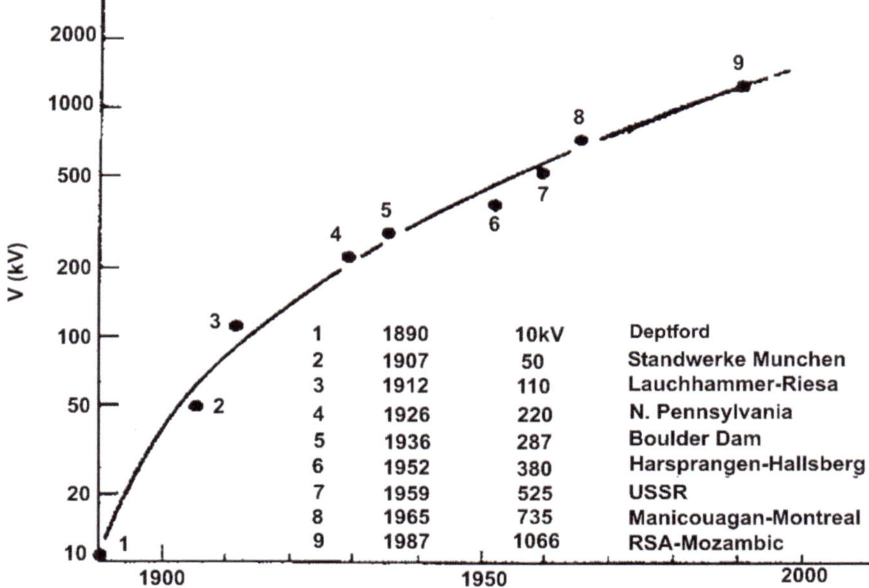

Fig. 6.2 Major AC transmission system in chronological order of their installations

Fig. 6.3 Photograph of AC OTL-1150 kV

$$V(kV)\quad 400\ 700\quad 1000\ 1200\ 1500$$
$$P(MV)\quad 640\ 2000\ 4000\ 5800\ 9000$$

The appearance of two AC OTL of different voltage classes of single-circuit and double-circuit design is shown in Figs. 6.3 and 6.4. Today, transmission voltages of 230 (220), 287, 345 (330), 500, 735 (750), and 765 kV are commonplace, with the first 1100 (1150)-kV line already energized in the early 1990s. (In parentheses are the nominal OTL voltages, adopted in Europe, including Russia.) The trend is motivated by economy of scale due to higher transmission capacities, more efficient use of right-of-way, lower transmission losses, and reduced environmental impact.

The two main requirements to transmission systems: (a) the interconnection of stations and neighboring systems to provide security, economic operation and the exchange of energy on a buy–sell basis, and (b) the transport of electrical energy from remote generation to load centers are met by selection of the most economical overhead line design, commensurate with the various constraints imposed by environmental and national considerations. The design and approval process for new lines can take many years; following public enquiries and judicial proceedings before planning, permission is ranted and line construction begins. Typical objections to new overhead lines, particularly in industrialized countries, are:

- visual deterioration of open country areas,
- possibility that electromagnetic field propagation may cause interference with television, radio, and telecommunications and with an increasing awareness of
- public health issues,
- emission of noise by corona discharges, particularly at deteriorating conductor
- surfaces, joints, and insulation surfaces,
- the danger to low-flying aircraft,
- a preference for alternative energy supply, such as gas and local environment-friendly power generating plants including solar cells and wind power turbines, or measures for reducing electric energy demand, such as the better thermal insulation in houses and commercial premises, lower energy lighting, natural ventilation in place of air conditioning, and even changes of lifestyle. Power system planners are required to show that extensive studies using a range of scenarios and sensitivities have been carried out, and that the most economical and least environmentally damaging design has been chosen. Impact statements are also required in many countries to address the concerns regarding many issues raised by local groups, planning authorities, and others.

Technically, the key issues, which have to be decided are:

- whether to use an overhead line or an underground cable,
- the siting of substations and their size required to contain the necessary equipment for control of voltage and power flow,
- provision for future expansion, for an increase in demand, and in particular for the likelihood of tee-off connections to new load centers,
- the availability of services and access to substation sites, including secure,

Fig. 6.4 Photograph of
double-circuit AC
OTL-400 kV

- communications for control and monitoring,
- whether to use a three-phase lines that are duplicated on one tower or mixed-voltage systems on a single tower [1].

Due to the rapid growth of the power grids and increasing requirements of consumers to the electricity quality energy management is improving. Energy management is the process of monitoring, coordinating, and controlling the generation, transmission, and distribution of electrical energy. The physical plant to be managed includes generating plants that produce energy fed through transformers to the high-voltage transmission network (grid), interconnecting generating plants, and load centers. Transmission lines terminate at substations that perform switching, voltage transformation, measurement, and control [2, 3].

The creation of United Electrical Power System (UEPS) in the USSR and others developed countries went in the following way:

- *End of the XIXth century—First quarter of the XXth century*—Unification of power plant for parallel working (at the beginning—on city scale, further—on regional scale).
- *First third of the XXth century*—Formation of *Interconnected Power Systems (IPS)* based on 110 and 220 kV overhead transmission line (OTL).
- *Second third of the XXth century*—Creation of the *Unified IPS* based on 330 and 500 kV OTL.
- *Last third of the XXth century*—unification of IPS into *Unified Power System (UPS)* by means of 500, 750 and 1150 kV OTL.
- *End of the XXth century—beginning of the XXIst century*—Restructuring and deregulation of the vertically integrated energy industry *(UPS)* to display it on the market.
- *Beginning of the XXIst century*—Development of the so-called *Strong Grid and Smart Grid.*

Further development (increase of scales) and perfection (improvement of quality) of systems of transportation and distribution of the electric energy depend on solving a number of problems:

- Planning of the perspective development of the EPS grid subsystem under conditions of highly uncertain initial conditions;
- Necessity of creation of high-power energy bridges caused by very large no uniformity of distribution of electric energy manufacturers and consumers in a particular country and on the global scale;
- Creation of steady-state grid subsystems in countries with high proportion of alternative electric energy sources integrated into the traditional power supply system;
- Spatial restrictions on building of transmission lines in cities and densely populated regions;
- Creation of electric interconnections of power supply systems with different nominal frequencies;
- Solution of complex scientific and technical problems associated with creation of active-adaptive electric grids and development and application of satellite information-measuring systems;
- Provision of high survivability of transmission lines under conditions of high activity of population and increased frequency and intensity of natural

Table 6.1 Lengths of the OTL different voltage classes in Russia (currently)

U (kV)	Extension, thousand km
500–750	30
220	140
110–150	281
35	187
3–20	1054
0.38	850
Total	2542

Fig. 6.5 Ice on the wire of OTL-110 kV

Fig. 6.6 OTL-220 kV support deterioration under the influence of a hurricane

perturbations (hurricanes, sharp temperature fluctuations, heavy snowfalls, etc.), Figs. 6.5 and 6.6.

They will be briefly discussed below.

6.2   Main Tendencies in the Development of Power Transmission Systems

Among specific features of the future global power engineering, will be centralization of energy distribution and a great diversity of generating sources. Its characteristic features are combination of large-sized powerful centralized and relatively small enterprises operating in a common grid, application of hybrid schemes of electric power and heat generation (cogeneration), compatibility of energy and production technologies with complete utilization of wastes and secondary resources, and formation of interconnected power systems. Known advantages of parallel operation of power plants led to intensive development of electric networks (grids) of higher voltage classes, expansion and consolidation of the electric power systems, and formation of large, geographically extensive, including interstate, power systems [4]. Positive experience on the creation and operation of large interconnected power systems in Western Europe, North America, the former USSR, and Eastern European countries has been accumulated.

There are all prerequisites for the development of the electric power engineering of the world economy in this direction.

The tendency observed in the past few decades, namely, formation of large national and international interconnected power systems is of particular importance. An important role in the formation of united power system in the Eurasian continent the is played by Russia having large fuel and energy resources and the largest in the world centralized interconnected power system called United Power Grid of Russia (UPGR)—Joint-Stock Company.

Before the formation of independent states from the USSR, there were three large interconnected power systems on the European continent: the UCPTE including 12 countries of Western Europe (Belgium, Germany, Spain, France, Greece, Italy, Yugoslavia, Luxembourg, Netherlands, Austria, Switzerland, and Portugal), the NORDEL System including four countries of Northern Europe (Norway, Denmark, Finland, and Sweden), and the Mir Power System including countries-members of the former Council for Mutual Economic Assistance. Asynchronously with the UCPTE, the Great Britain Power System operated through a dc cable.

The installed capability of electric power stations included in the UCPTE amounted to more than 390 million kW; it amounted to 85 million kW for the NORDEL System and to more than 400 million kW for the Mir Power System. The Mir Power System was connected to the UCPTE through three DC inserts having net capability of 1750 MW and with the NORDEL by a dc insert having a capability of 1100 MW. Power systems of Eastern European countries and the UPG of the former USSR were connected by three 750-kV overhead transmission lines (OTL), four 400-kV OTL, and four 220-kV OTL to deliver electric power from the USSR to Eastern European countries. The energy supply amounted to ~ 40 billion kWh in separate years.

The integration processes in the UCPTE and NORDEL System are intensified. In 1994, a dc cable line between Switzerland and Germany having a length of about 250 km and capability of 600 MW was put into service. Two projects of connecting the Norway Power Grid to the continental Europe are being considered. The first project is aimed at interconnection of Norway and Germany power systems, and the second project is aimed at interconnection of Norway and Holland power systems. The feasibility of building an interconnecting dc transmission line between Sweden and Poland was justified. The feasibility of connecting Latvia, Lithuania, and Estonia power systems to the NORDEL System and UCPTE will be examined further.

In 1994, the electric power exchange in the UPCTE, including the third countries, amounted to 155.9 billion kWh or 10% of the net electric power generated by UCPTE countries, and the corresponding exchange in the NORDEL System amounted to 39.3 billion kWh or 11.2%.

In the Mir System, disintegration processes started just after disintegration of the USSR, and the mutually advantageous electric power exchange inside the system significantly reduced. In October 1995, the CENTREL interconnected power system including Hungary, Czech, Slovakia, and Poland interconnected power systems and the interconnected power system of the eastern part of Germany joined the UCPTE. The installed capability of the extended UCPTE amounted to more than 470 million kW. The integration of Bulgaria and Romania power systems to the UCPTE is also planned. At the end of September—beginning of October 1995, the Bulgaria power system was disconnected from the Ukraine UPG and was switched to synchronous operation with Romania, Greece, Albania, and the former Soviet Federal Republic Yugoslavia power systems. This experiment was considered as a step toward connection of power systems of Southern European countries to the UCPTE. Turkey is the next candidate for connection to the UCPTE. The feasibility of interconnection of the Turkey power system with power systems of countries of the Mashreq economic zone (from Syria to Egypt) is studied. After commissioning a deep-sea ac cable connecting Spain and Morocco, power systems of Morocco, Algeria, Tunis, and Libya (countries of the Maghreb zone) joined the UCPTE. The feasibility of interconnection of power systems of Mashreq and Maghreb countries is studied. Thus, a large interconnected power system of countries of the Mediterranean Sea basin is being built. This interconnected power system will operate in parallel with the UCPTE. It is also planned to study the feasibility of joint operation of the Turkey power system with power systems of Trans-Caucasian republics including Armenia, Georgia, and Azerbaijan.

At the same time, the UPGR continues to operate synchronously with power systems of Baltic countries, Byelorussia, Ukraine, Moldova, and Kazakhstan. Power systems of Azerbaijan, Armenia and Georgia have retained the feasibility of synchronous operation with the UPGR.

Under these conditions, a central problem in cooperation of countries of the European continent in the field of electric power engineering has become the use of already existing 11 OTL connecting the Commonwealth Independent States (CIS) and Eastern European countries in the building of which large sums of money were invested. Different ways of further development of this cooperation are suggested. One of the variants envisages the transfer of dc inserts to the boundaries of CIS and Eastern European countries.

The best way of cooperation on the Eurasian continent is the creation of a common electric power market as a basis for an interconnected power system. A number of international projects are aimed at solving this problem.

Baltic Electric Power Generating Ring. This project is aimed at creating a high-power grid connecting power systems of 11 Baltic countries including Denmark, Sweden, Norway, Finland, Russia, Estonia, Latvia, Lithuania, Byelorussia, Poland, and Germany, Fig. 6.7. Another project of the East–West

Fig. 6.7 Baltic electric power generating ring

Power Bridge envisages the building of a 4000-MW dc transmission line connecting power systems of Russia, Byelorussia, Poland, and Germany and is conceptually a part of the first project.

It is assumed that the Baltic Ring will allow the operation of power systems of participating countries to be improved and as a whole, will foster the economic development of Baltic countries.

A meeting of 17 electric power-generating companies from 11 countries of this region devoted to the creation of the Baltic Electric Power Generating Ring was held.

It should be noted that positive experience on cooperation of the UPGR with the NORDEL System has already been accumulated. Work is underway on an increase in the capability of the dc insert with Finland up to 1400 and then to 2000 MW. The feasibility of connection of the Karelian and Kola power systems with power systems of countries included in the NORDEL is being studied.

Black Sea Interconnected Power System. The majority of countries involved in the Black Sea Economic Community (BSEC) including Ukraine, Romania, and Bulgaria supported the proposal of the JSC UPGR to create the BSEC United Power Grid. Its formation is aimed at interconnection of regional power systems into the high-power grids; some of them already exist. Such an interconnected power system will allow BSEC countries to develop electric power engineering in the entire region in the best way, to use rationally the available power resources, to increase reliability of electric power supply to customers, to exchange electric power to mutual advantage, and to affect positively the economic development of all countries in the region. High-voltage grids built by countries-members of the Council for Mutual Economic Assistance will provide the basis for the interconnected power system, including 400- and 750-kV power grids connecting Russia, Ukraine, Moldova, Bulgaria, and Romania in the southwest, 330- and 500-kV power grids connecting Russia, Georgia, Armenia, and Azerbaijan in the southeast, and 220-kV OTL connecting Trans-Caucasian countries and Turkey.

Other projects of interconnected power systems. Variants of the connection of power systems of Central Asia and Iran and Turkey are being considered, and problems of integration of interconnected power systems of Russia and China, Japan, and Korea as well as Russia and USA are being studied.

Electric power engineering of China develops rapidly; the annual increment of electric power generation amounts to 7–9%. The net annual electric power generation in China exceeds 900 billion kWh. China is interested in the electric power transfer from Russia. Potential sources of electric power for export are located in Siberia (Boguchansk, Bratsk, and Ust-Ilim HPS and Berezovskii State District Power Station (SDPS)) and Far East (an APS in Khabarovsk Region, HPS and TPS in the Amur Region and Yakutia, and a tidal power station in the south of the Okhotsk Sea). 500-kV ac OTL with dc inserts having a dc carrying capacity of 1.5–2 million kW can be used to transfer electric power. In the East interconnected power systems, Amur, Khabarovsk, and Far East power systems are considered as transmitting one. OTL at voltages up to 500 kV inclusively can be used to export electric power.

The main premises for electric power import by Japan consist in the absence of its own fuel and energy resources and extremely high population density. Sakhalin TPS burning a shelf gas or South-Sakhalin coal, HPS and APS of the Far Eastern Interconnected Power System, and a tidal power station in the south of the Okhotsk Sea can be considered as potential sources to export electric power from Russia to Japan. For this aim, transmission lines can be built through Sakhalin Island and two shallow and narrow channels (Tatarskii and Laperuza) or through the territories of China and Korea and the Korean Channel 200 km wide.

With allowance for long transmission lines, the electric power transfer to the USA is planned in small amount if the main expenses on building of the OTL through the Bering channel and mastering the hard-to-reach coastal zone of this channel will be covered by the headquarters of building of transcontinental railway through the Bering channel.

The implementation of the above-considered international projects as well as of the suggested variants of interconnected power systems will allow the Japan–China–Siberia–Kazakhstan–European part of Russia–Western Europe powerful extended electric power system to be formed and will be an important stage in the creation of the Interconnected Power System on the Eurasian continent with the net electric power of the order of 60% of the electric power of all power stations in the world in which the UPGR, by virtue of its geopolitical position, can become a central link.

The required transfer capability of intersystem connections can be estimated based on the practical recommendations of the UPG of the former USSR according to which the net transfer capability of intersystem connection in cross sections dividing the powerful interconnected power system into two parts should be of the order of 2–3% of the maximum load of the smaller part of the examined interconnected power system. With allowance for this condition, the required transfer capability of intersystem connections for the Eurasian Interconnected Power System in the territory of Russia and Kazakhstan should exceed 10 GW. Such transfer capability can be obtained only with the use of the super-high voltage (1150 kV of ac and 1500 kV of dc) transmission lines.

The North American transmission system is interconnected into a large power grid known as the North American Power Systems Interconnection. The grid is divided into several pools. The pools consist of several neighboring utilities, which operate jointly to schedule generation in a cost-effective manner. A privately regulated organization called the North American Electric Reliability Council (NERC) is responsible for maintaining system standards and reliability. NERC works cooperatively with every provider and distributor of power to ensure reliability. NERC coordinates its efforts with FERC as well as other organizations such as the Edison Electric Institute. Figure 6.8 shows a diagram of high-voltage transmission lines of the United States.

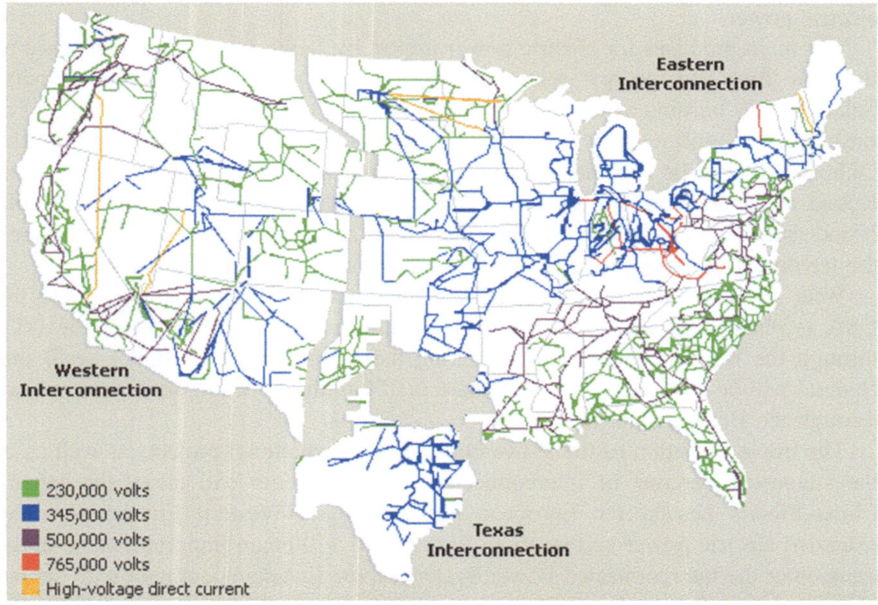

Fig. 6.8 Diagram of high-voltage transmission lines of the United States

NERC currently has four distinct electrically separated areas. These areas are the Electric Reliability Council of Texas (ERCOT), the Western States Coordination Council (WSCC), the Eastern Interconnect, which includes all the states and provinces of Canada east of the Rocky Mountains (excluding Texas), and Hydro-Quebec. These electrically separate areas exchange with each other but are not synchronized electrically.

The electric power industry in the US is undergoing fundamental changes since the deregulation of the telecommunication, gas, and other industries. The generation business is rapidly becoming market-driven. The power industry was, until the last decade, characterized by larger, vertically integrated entities. The advent of open transmission access has resulted in wholesale and retail markets.

Trends in the development of electric power systems of the USA united at the state, interstate levels formed in the second half of the XXth century acquired clear-cut temporal, and territorial guidelines to which the world electric power engineering adheres in the new century.

Nowadays interstate electric energy associations operate in the Western and Central Europe (Union for the Coordination of Transmission of Electricity—UCTE), Northern Europe (NORDEL), North America, CIS countries, and so on. The EPS integration and formation of large electric energy associations are underway in Asia, Africa, and Southern and Central America.

The Trance-European Synchronously Interconnected System (TESIS) is formed by Trans-European Western, Central, and South-Eastern European countries. Further development of this electric energy association can proceed in collaboration with CIS electric energy association (established in 1992) based on intensification of interconnections on alternating current and construction of electricity transmission lines on direct current within the limits of widely discussed complex projects "Baltic Ring," "Direct Current Energy Bridge," and "East–West." Analogous way of development of high-power electric grids is characteristic for energy associations of North America.

Energy association of the countries of Central Asia (Southern Kazakhstan, Uzbekistan, Kyrgyzstan, Tajikistan, and Turkmenistan) continues to work successfully. In future intensification of its interconnections with the United Energy System (UES) of Russia and Southern countries is expected. India intensively develops electric power engineering and is in the way of creation of the national energy association and development of its integration with neighboring countries. Interstate energy association is formed on the Indochina peninsula.

One of the promising regions from the viewpoint of creation and development of interstate energy associations is Northeast Asia including Eastern Siberia and Far East of Russia, Mongolia, China, North Korea, South Korea, and Japan. For this, there are essential prerequisites associated with different territorial deployment of

Fig. 6.9 Main Grid of the USSR and Russia: OTL-220, 330, 500, 750, and 1150

energy resources and centers of energy consumption as well as with essential potential system effects caused by the formation of interstate energy associations. The infrastructure of the main electrical grid in this region is also developed.

Russia is interested in the formation of united energetic and energy transport infrastructure in adjacent regions of Europe and Asia, in the development of international energy transmission systems including electric grids. For these purposes, the state encourages participation of the Russian joint-stock companies and firms in working out and implementation of large-scale international projects on the creation of electric energy associations "Baltic Ring," "United Energy System of Caspian and Black Sea Countries" and energy bridges "East–West," "Siberia–China," "Sakhalin–Hokkaido," etc., Figs. 6.9 and 6.10.

For several years, experts in energy and politics have been considering various versions for creating the Asian Energy Super Ring.

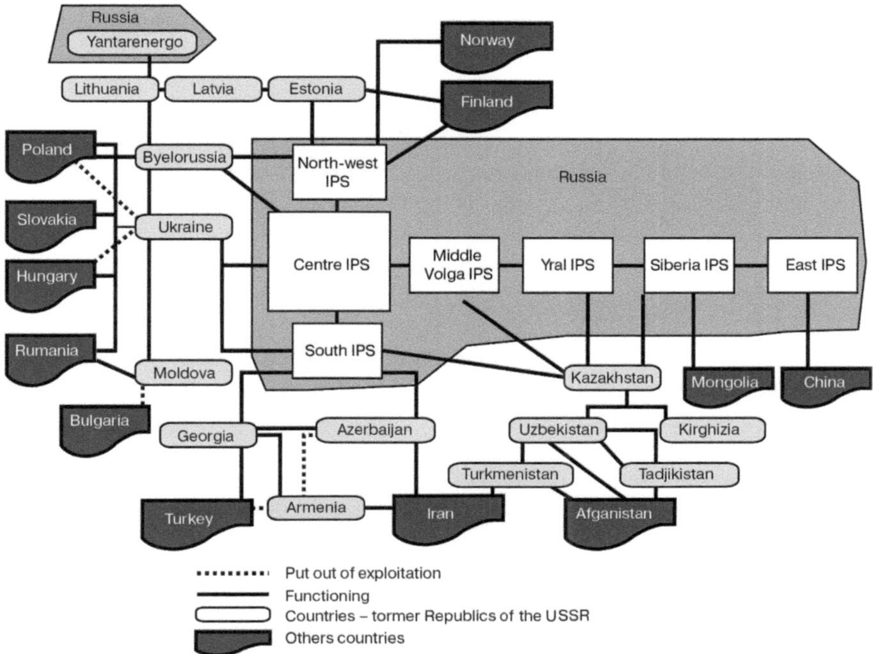

Fig. 6.10 Interstate United Energy System (UES) of Russia (as of 2010)

6.3 Technical Problems of the Electric Grid Complex

6.3.1 Provision of Uninterrupted Electric Grids

The uninterrupted electric power supply of consumers by qualitative electric power is determined by the ***reliable*** work of the energy system that requires a successful solution of problems listed in Sect. 6.1.

Their solution must provide:

- required (standard) margin of static stability of the transmitted power in normal and post emergency regimes;
- required (standard) margin of static stability on the voltage of load units of the grid;
- heat resistance of grid elements (lines, transformers, switches, etc.) in normal and post emergency regimes;
- voltage levels admissible for the equipment of a power supply system;
- necessary operative margin of power to maintain the development of the competitive market of electric energy and power;
- decrease in electric power losses during energy transmission;
- maintenance of high-quality electric power.

The parameter **reliability** includes such EPS properties as survivability, non-failure operation, durability, repair ability, stability, and controllability of the regime [5].

Reliability has special definitions, which differ from the usual planning or operating usage. A relay can misoperate in two ways: it can fail to operate when it is required to do so, or it can operate when it is not required or desirable to do so. To cover both situations, there are two components in definition reliability:

Dependability—which refers to the certainty that a relay will respond correctly for all faults for which it is designed and applied to operate;

Security—which is the measure that a relay will not operate incorrectly in case of any fault.

Most relays and relay schemes are designed to be dependable since the system itself is robust enough to withstand an incorrect trip out (loss of security), whereas a failure to trip (loss of dependability) can be catastrophic regarding system performance [6].

Continuous increase in capacity and the expansion of territories served, as well as the emergence of new technologies for the production, transmission, and distribution of electricity increase demands on the reliability of the protection of the electric transmission lines. Each electrical element will have problems unique to itself, but the concepts of reliability, selectivity, local and remote backup, zones of protection, coordination and speed that may be present in the protection of one or more other electrical apparatus are all present in the considerations surrounding transmission line protection.

Since grids are also the links to adjacent lines or connected equipment, their protection must be compatible with the protection of all of these other elements. This requires coordination of settings, operating times, and characteristics.

The purpose of electric power system protection is to detect faults or abnormal operating conditions and to initiate corrective action. Relays must be able to evaluate a wide variety of parameters to establish that corrective action is required. Obviously, a relay cannot prevent the fault. Its primary purpose is to detect the fault and take the suitable action to minimize the damage to the equipment or the system. The most common parameters, which reflect the presence of a fault, are the voltages and currents at the terminals of the protected apparatus or the appropriate zone boundaries. The fundamental problem in electric power system protection is to define the quantities that can differentiate between normal and abnormal conditions. This problem is compounded by the fact that "normal" in the present sense means outside the zone of protection. This aspect, which is of the greatest significance in designing a secure relaying system, dominates the design of all protection systems [7].

Improvement of electric power system parameters of the developed countries also involved such elements, as supports and wires of overhead transmission lines (OTL). Works are underway on construction and putting into practice of OTL towers and aesthetic supports. The towers allow long spans to be designed for crossing water, woody, and boggy areas. In this case, the height of the overhead phase conductor exceeds 40 m, and the total height of the design reaches 60 m. The aesthetic supports should be graceful, aerial, and fit harmoniously into the surrounding landscape as independent design objects adapted as much as possible to aesthetics of the environment from near and far observation points.

Creation of new wires is intended to meet increasing requirements—high mechanical strength, thermal stability, and low ohmic resistance.

In the last quarter of the 20th century (up to now), increase is observed of the share of underground distribution transmission lines (UDTL). All-underground construction—widely used for decades in cities—now appears in more places. UDTL is much more hidden from view than OTL, and is more reliable. Cables, connectors, and installation equipment have advanced considerably making UDTL not so expensive. One of the main applications of UDTL is underground residential distribution and underground branches or loops supplying residential neighborhoods. The use of underground construction is benefit for mounting substation exits and creating systems for serving industrial or commercial customers, for crossing river, highway, or OTL.

Underground construction is expensive, and costs vary widely (from 25 to 1000 doll/m). The main factors that influence UDTL costs are:

(1) Degree of development—roads, driveways, sidewalks, and underground utilities; these and other obstacles slow down construction and increase costs;
(2) Soil condition—rocks and frozen ground increases time and cost of cabling.
(3) Nature of the terrain on which the cable will be laid: urban construction is more difficult not only because of concrete, but also because of traffic; rural construction is generally the least expensive per length, but also lengths, as a rule, are much larger.
(4) Method of laying the cable: concrete-encased ducts cost more than direct-buried conduits, which cost more than preassembled flexible conduits, which cost more than directly buried cable with no conduits.

Transmission cable lines are especially attractive for overcoming the big water obstacles (sees, big lakes and straits).

The choice of ways of solving the problem *controllability* of EPS is based on results of studying information about such properties of EPS: observability, predictability, and identifiability. Theory of transient processes and EPS stability has been developed and continues to be improved based of general theory of dynamic systems.

To provide reliability and efficiency of functioning of EPS and electric power market within the limits of the preset reliability level, preservation and strengthening of the centralized hierarchical system of real-time dispatch control and automated systems of dispatch EPS monitoring that have been developed in many countries (the USA, the USSR, Germany, France and some others) by the end of the last century have special importance.

It should be emphasized that the unconditional fulfillment of dispatch discipline have assumed great importance in connection with the accelerated development of a distributed (decentralized) energy sector.

It becomes more and more obvious that in the nearest future traditional approaches to the development of electric power grids will not meet new requirements on reliability, quality, and efficiency of electric power supply. A radical solution of these problems is provided by the new concept of EPS transformation into intellectual systems according to concept "Smart Grid" (see Sect. 6.3).

6.3.2   Minimization of Electric Power Transmission Losses

Electric power is different from other products in that it is transported at the expense of spending part of the product itself and, therefore, power loss is inevitable. In this regard, only a fraction of the electric power, which does not arrive to the consumer from the producer, can be attributed to the **"net loss" or "real loss"**. The other part reasonably is called **"technological power consumption for transmission through electric grids."** Below, for brevity, all the electricity not supplied to the consumer, is called "losses" (Fig. 6.11).

In electric grids of electric power and electrification enterprises with voltage of 220 kV and lower, the electric power losses make 78% of the total losses, including 28% in 110–220 kV grids, 16% in 35 kV grids, and 34% in 0.38–10 kV grids.

The electric power losses independent of loading (*conditionally constant losses*) make 24.7%, and the *loading losses* (dependent on the electric power transmitted through the grid) make 75.3% of the total losses. The loading losses involve 86% of

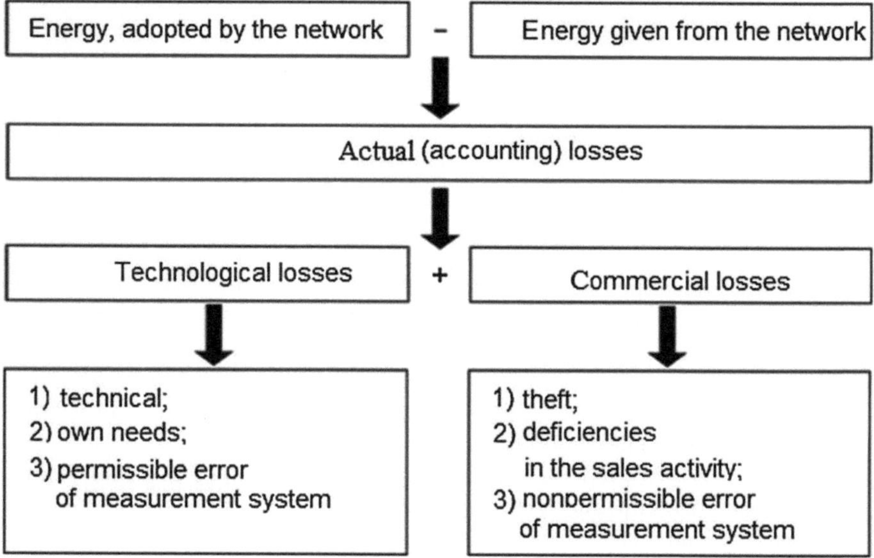

Fig. 6.11 The structure of electricity losses on the transmission stages and distribution

Table 6.2 Level of transmission losses in a number of leading countries and the global average

Countries and their associations	Electricity losses during transmission and distribution (%)
India	25
Mexico	16.2
Brazil	16.6
Russia	12.0
China	6.7
EU-27	6.7
USA	6.2
Canada	7.3
Japan	4.6
The whole world	*8.8*

losses in power transmission lines (PTL) and 14% of losses in transformers. The conditionally constant electric power losses involve 67% for idling transformers, 11% for internal needs of substations, and 22% for other losses.

The level of electricity losses during transportation in different countries lies in a wide range—from 4.6 to 16.6%, Table 6.2.

According to international experts, an amount of electricity losses during transmission and distribution can be reduced to an economically justified and technically achievable value—4–5% (a value close to the limit determined by physical laws). The potential for reducing losses in the grids (real today) exceeds 20% of the total losses.

The reduction of excessive power losses in grids in addition to limiting the growth in electricity tariffs can be an important source of investment (along with other sources) into the electric power systems for a radical improvement of their technical condition.

The study of the dynamics and structure of the electricity losses have revealed a number of important conformities:

1. The transition from centralized (state) to market mechanisms of energy management is accompanied by an increase in electricity losses in grids. The effect was observed not only in the Russian electric power system during and after the 1998–2008 reform, but in other countries that went to the privatization and decentralization of this sector. The main reason—the weakening of the institutional/executive and technological discipline.
2. The relative power losses in the grids the lower, the higher is the share of industrial consumption in the productive supply.
3. Inclusion of the loss ratios in tariff for electricity transmission leads to a dangerous trend fit these standards under the actual losses.

Each of this conformity includes dozens of items.

The set of measures to reduce technical losses of electric energy in an enlarged form is shown in Fig. 6.12.

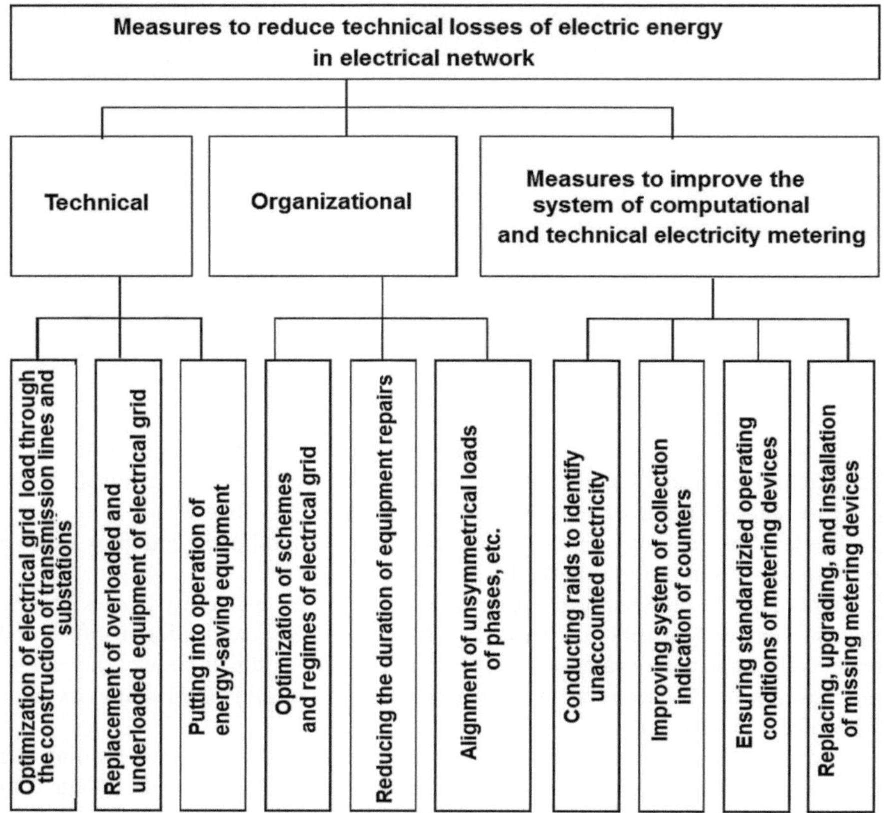

Fig. 6.12 Measures to reduce technical and commercial electricity losses

Reduction of power losses in grids is a difficult complex problem. The problem solution depends on information, methodical, mathematical, logistical and organizational support.

Numerous and diverse technical/technological and organizational measures to reduce the losses of energy in grid can be represented in the form of 4 groups:

(1) energy-saving solutions in the construction, reconstruction and development of electricity networks, putting into operation energy-saving equipment;
(2) optimization of circuit topology and regime parameters in operation and operational management of electricity networks;
(3) improvement of calculations of loss ratios, power balances feeders, power centers, and electric networks in general;
(4) training and motivation of staff to reduce losses, control of its efficiency, improving the organization of work.

Problems of energy losses in electrical networks considered in textbook in detail in [8]. Here we confine ourselves only brief remarks regarding the reduction of losses of electricity in transport and the maintenance of its standard quality by reactive power compensation.

The most simple, but difficult feasible way to reduce electricity losses during transmission is reduction of electric resistance of conductors and/or the current density.

There are several ways to reduce the resistance of a wire:

1. To use for an electricity supply all the line (Not advisable to have a backup line);
2. To reduce the resistance of the lines and the extent of the increase in cross-section of cable and overhead lines, as well as to use conductors with a lower specific resistivity;
3. To design a network with a minimum of contact connections, and to perform a junction of the cable cores and conductors with the use of special clamps and couplings to ensure minimum contact resistance.

6.3.2.1   Increasing the Power Factor

Increasing the power factor today is the more widespread and effective method of energy loss decrease during transmits. Power factor is the ratio of the effective power in kilowatts to the apparent power in kilovolt-amperes ($\cos\varphi$). The latter is a geometric sum of the active power that performs active work due to the use of energy and the reactive power (RP) that provides magnetic and electric fields in electrical equipment and power exchange between them necessary for a number of power consumers.

By means of the choice of an economically viable cross section of conductors (wires) and reactive current compensation (equalization of its inductive and capacitive components), it is possible to increase the power factor (in the limit to one) and to reduce the voltage drop in the transmission line and power losses (Table 6.3).

The voltage drop and power loss in the conductors are determined by the total current, comprising the active and reactive components, and are given by the formula

Table 6.3 Impact of losses on the value of $\cos\varphi$

Previous $\cos\varphi$	0.5	0.5	0.6	0.6	0.7	0.7	0.8
New $\cos\varphi$	0.8	0.9	0.8	0.9	0.8	0.9	0.9
Current decrease (%)	37.5	44.5	25	33	12.5	22	11
Decrease the losses according with resistance (%)	61	69	43.5	55.5	23	39.5	21

$$I = \frac{\sqrt{P^2 + Q^2}}{U\sqrt{3}}, \tag{6.1}$$

where P and Q are active and reactive power transmitted on the line and U is the phase voltage at the end of the line.

Given that the ratio of electric power

$$E = \cos\varphi = \frac{P}{\sqrt{P^2 + Q^2}}, \tag{6.2}$$

Get $\Delta P = P^2 r_0 / (3U^2 \cos^2 \varphi)$, that is, with an increase in the power factor, losses in the power lines decrease. This also reduces the voltage drop, since it is inversely proportional to the power factor. Thus, it is expedient to pass only to the energy corresponding to the resistive load of the consumer.

Full power is calculated as a geometric sum of active and reactive powers. It determines the total power requirements for generating, grid, and distribution structures—all network equipment (generators, transformers, transmission lines, distribution systems, etc.) have to rely on the power, significantly exceeding the useful (active power). Energy companies face additional costs for equipment and additional energy losses, forcing them to charge customers an additional fee in cases where the proportion of reactive power exceeds a certain threshold value.

Usually, as the threshold value, $\cos\varphi$ is chosen not less than 0.9. In practice, $\cos\varphi$ after reactive power compensation is in the range 0.93–0.99. In the aspect of reducing the losses of energy in the lines, it is desirable fully compensate for the reactive power, but this cannot be realized, because the implementation of such a solution requires the installation of special compensating devices with costly RP regulation.

For each node of the power system the so-called *technical condition of reactive power compensation* must be carried out, namely RP balance at any node of grid, breach of which varies the voltage at this node. From the viewpoint of maintaining the necessary voltage mode by the consumers, it is unacceptable deficit of RP.

In addition to technical specifications, there are technical and economic conditions that reduce the flows of RP, which lies in the fact that in the RP is compensated directly by consumers:

(1) reducing the current transmission network elements, allowing to reduce the cross-section of wires:

$$I = \frac{\sqrt{P^2 + (Q_i + Q_{e.d})^2}}{U},$$

(2) reducing the total power, which reduces the power transformers and their number:

$$S = \sqrt{P^2 + (Q_i + Q_{e.d})^2},$$

(3) reducing active power losses ΔP, thereby making it possible to reduce the power generators in power plants;

(4) reducing the reactive power losses ΔQ, allowing to reduce the power compensating devices;

(5) reducing active energy losses $\Delta W = \Delta P \tau$, allowing to save fuel in power plants.

Industrial companies feeding power system set the economic value of reactive power Q_e, it may refer to the period of peak grid load. Knowing the reactive load $Q_{ent.}$ of an enterprise or its maximum load P_{max}, it is possible to determine the capacity of compensating devices $Q_{cd.}$ to be installed on industrial enterprise:

$$Q_{cd.} = Q_{ent.} - Q_e = P_{max}(tang\varphi_{ent.} - tang\varphi_e) \tag{6.3}$$

where tang φ_{ent}—the actual ratio of the reactive load of the enterprise; tang φ_e is the reactive load factor corresponding to Q_e.

6.3.2.2 Regulation of RP in Power Systems

In power systems, RP sources are used in the grids of 110 kV and above for the following tasks:

- reduction of active power losses and power energy,
- load voltage regulation in the nodes,
- increase in the capacity of electric power,
- increase in the stock of the static stability of electric power transmission lines and generators of power stations,
- increase of the dynamic stability of power transmissions,
- surge suppressor,
- balancing mode.

Employed in the device of targeting RP balance (RP sources) are divided into two radically different groups: (1) synchronous machines (power generators, compensators, motors), and (2) static RP sources or static reactive power compensators (capacitor banks, reactor—based transducer device (rectifiers, inverters) with artificially switched thyristors or combinations thereof). The main parameter of RP source regulation is the voltage at the point of its connection or reactive power load; to regulate for reactive power load, RP source or a combination of both can be used.

Sources of RP of *the first group* allow fluently adjust the RP in regime of generation and consumption (in terms of "production" and "consumption" of the RP there is a certain conventionality, but in doing so, is emphasized that the interaction of capacitive and inductive elements in the grid has compensating character).

Sources of RP of *the second group* allow one to regulate the RP only stepwise. For their switching in networks with voltage up to 1 kV, conventional contactors are used, in networks with voltage greater than 6 kV, circuit breakers or thyristor keys are used.

RP consuming by reactors can be controlled both stepwise using the same method as for capacitors and switching equipment and fluently using thyristors. Special group includes the swinging reactors capable to change of fluently consumption of RP parametrically (without regulator) depending on a voltage applied at the connection point.

For RP compensation, improving the manageability of the modes large electric power systems use modern means of regulation reactive power compensation: block of static compensators (BSC), static thyristor compensators (STC), shunt reactor (SR), controlled shunt reactor (CSR), automated control system (ACS), static compensator (STATCOM), Fig. 6.13, and vacuum-reactor groups (VRG).

Creating these efficient devices and circuits for their switch on was made possible due to the rapid development of power electronics in the last decades.

As an example, in Fig. 6.14 is shown the scheme of parallel compensation based on capacitors and a thyristor-controlled reactor: (a) circuit and (b) current and voltage waveforms [9].

Fig. 6.13 A static synchronous compensator

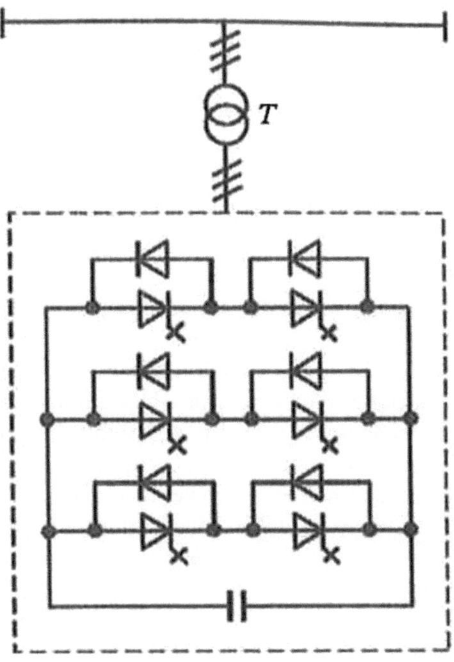

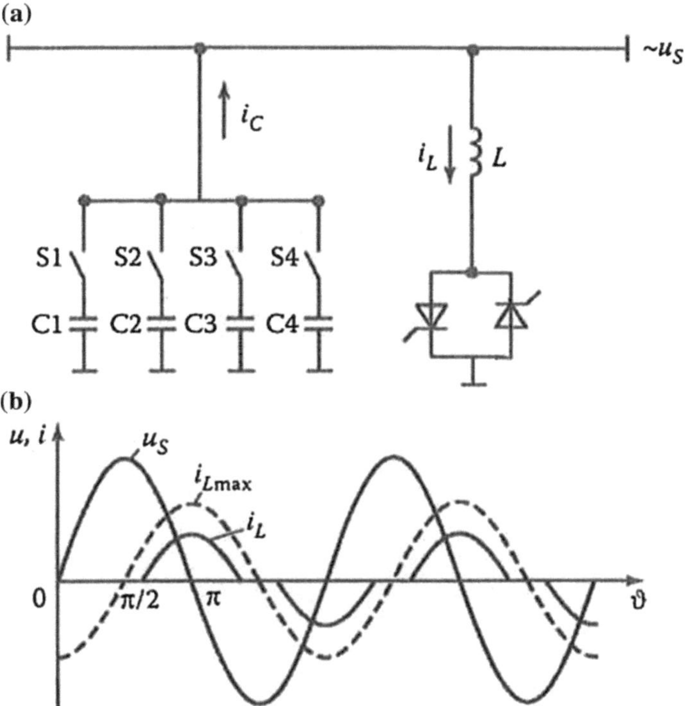

Fig. 6.14 Scheme of parallel compensation on the basis of capacitors and a thyristor- controlled reactor: **a** circuit and **b** current and voltage waveforms

Combined use of capacitors and a reactor with thyristor control permits smooth regulation of the reactive power, whether it is capacitive or inductive. Note that switches S1–4 turn on the capacitors at zero voltage. For the thyristor compensator, the range of reactive—power regulation is

$$\frac{U^2 \cdot w \cdot C_\Sigma}{U^2/w \cdot L_\Sigma}$$

where C and L are the total capacitance in the line and the reactor inductance, respectively; W is the angular frequency of the grid; and U the effective grid voltage. This device is not only fast, but also reliable.

Various versions of the device are currently in use. An alternative to series thyristors is the biased reactor. Its benefit is that only low power is required to control the bias current. This is usually a low-voltage, low current device. It also permits modification of the compensating power by regulating the inductance.

Two main circuits are used for series capacitor compensation based on thyristors, Fig. 6.15.

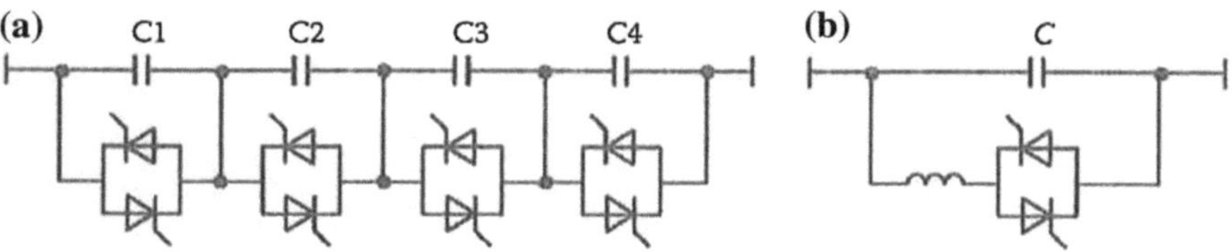

Fig. 6.15 Series capacitive compensation: **a** the thyristor-switched capacitors and **b** the thyristor-controlled circuit

First, the capacitive sections are shunted by antiparallel thyristors in series, Fig. 6.15a. To eliminate malfunctioning of the semiconducting switches under the action of the capacitors discharge currents, these devices are switched on when the voltage in the corresponding capacitor section passes through zero. Second, the capacitors and reactors are connected in series with antiparallel thyristors, Fig. 6.15b. The reactor operates analogously in the system with parallel compensation. The current is regulated by phase control of the thyristors. Depending on the thyristor control angle, the reactors compensate the series capacitance in the line. In the absence of capacitor compensation, the reactors are switched on to shunt the capacitors and are switched off in total compensation.

Additional expenditures for the means of regulated compensation in the backbone network (compared to uncontrolled) are justified not only by reducing the loss of energy. Their use can be justified when the additional positive effect is reached—an increase in the stability of the power and stability of the load, as well as increase of the power quality in the network.

6.3.2.3 Regulation of RP in the Power Supply Systems with Nonlinear Load

In industrial power supply systems, RP sources are used to compensate for the RP consumed by sharply variable powerful load balancing. To do this, the RP sources are used that can generate it—synchronous machines and capacitor banks. However, the former, allowing one smoothly adjust the RP; have a high inertia due to the time constant of the excitation system, which is a disadvantage. On the contrary, the capacitor battery (CB), especially switched thyristors, have a high speed (10–20 ms) with a stepped regulating PM. In those cases (for example, providing static of power stability) when the step regulation is unacceptable, a combination of RP sources is used, which enables one to smoothly adjust the PM at high speed. They usually consist of a stepwise adjustable CB and smoothly controlled reactor connected in parallel.

In *power supply systems with nonlinear (non-sinusoidal) load* generating harmonic currents, RP sources can function as filter-compensating devices.

To reduce the consumption of power by the RP, it is advisable to carry out activities that do not require installation of special compensating devices, as in the production of reactive energy, where compensation devices consumed active energy efficiently. These activities include:

(1) increasing the download process units and keeping it as constant as possible over time, which is accompanied by an increase in the load factor of electric motors and $\cos\varphi$, while reducing up to 40% the power consumed by asynchronous motor; supply voltage reduction of switching windings of a star with a triangle reduces engine power by 3 times;

(2) use of thyristor voltage regulators motor to control signals from the sensor $\cos\varphi$ and other technical regimes of electric motors and transformers;

(3) replacement, rearrangement, and disconnection transformers downloaded on average by less than 30% of their rated power. Particular attention should be given to the automation of two-transformer substations: when the load of transformers is reduced below 35%, they must be switched off for this period without affecting the automatic transfer switch;

(4) replacement of low-loaded engines by engines by engines of smaller capacity;

(5) replacement of induction motors and their synchronous application in all new installations of the drive where it is acceptable for technical and economic reasons;

(6) automatic control of excitation of synchronous motors and if necessary— switch-capacitor banks.

Means for $\cos\varphi$ adjustment are most effective if they are used in close proximity to the load. Since $\cos\varphi$ may vary over time due to changes in the characteristics and composition of the equipment, its adjustment must be performed at regular intervals from 3 to 10 years (depending on the nature of the enterprise, parameters, and modes of operation of the equipment).

6.3.2.4 Decrease of Energy Losses in Transformers

Energy losses in transformers are divided into two main types—conditional permanent loss or "iron losses" (i.e. in the core) and load losses or "copper losses" (i.e. in the windings). The "iron losses" are proportional to U^2 and about 0.2–0.5% of the rated power of the transformer. They can be reduced due to a decrease of the thickness of the electrical steel sheets and using higher quality steel. "Copper losses"—the omics losses proportional to I^2 are about 3.1% of the rated power of the transformer (when fully loaded).

The following ways are used to increase the efficiency of the transformer substations:

- if the power consumed by the load is less than 40–50% of the nominal (P_n), the energy-saving measures are advisable to disable one or more transformers and thereby to bring the load to other optimal values;
- if the power consumption of the load exceeds 75% P_n, then the efficiency of transformers can only be optimized by the installation of additional capacity;

overcapacity creates conditions for uninterrupted operation in the event of failure of one or more transformers;
- modernization of the transformer substations or only replacement of worn-out transformers is preferred to installation of transformers with low loss level, which allows them to be reduced by 20–60%; in this case, the payback period, as a rule, is only about a year.

Improvement of the parameters and modes of operation of transformers in addition to reducing the "iron losses" and "copper losses" is also focused on ensuring the efficiency of the network by maintaining appropriate quality electricity supplied to consumers, and regulation of load voltage.

6.3.3 Power Quality

Production processes largely depend on the quality of electric power. In general, the low quality of electricity can be described as changes in power supply, leading to disruptions of the production process or damage to electrical equipment. According to some experts, over the next 20 years the quality of power supply will be the biggest problem in the power industry, translated into most sectors of the economy. The basis for achieving a high quality of electricity are three components: production of high-quality electricity, uninterrupted transmission, and distribution of reliable networks [10].

The damage caused by non-compliance with regulatory requirements for power quality parameters can be subdivided into two parts: technological and electromagnetic components.

Technological component damage manifested in the reduction of electrical performance and manufacturing process that may increase the energy consumption due to the increase in the duration of the process or a mass marriage of products, reduce the efficiency of the equipment, etc. On energy, efficiency primarily affects technological component damage. Accordingly, as the level of energy efficiency associated with the power quality can be accepted optimal power mode, providing maintenance of a particular group power quality index at an acceptable level. Determining the number of them depends on the sensitivity to them of specific consumers.

Electromagnetic component damage is associated with reduced life of power equipment in connection with its accelerated wear, with an increase in the measurement error of the instrument electrical quantities and metering of electricity, with the failure of automation systems, relay protection, a shutdown or failure of the capacitor system as a result of resonance phenomena at higher harmonics, with increased levels of power and energy losses, etc.

It is clear that there can be completely different definitions for power quality, depending on frame of reference. For example, an energy company may define power quality as reliability and show statistics demonstrating that its system is

99.98% reliable. Criteria established by regulatory agencies are usually in this vein. A manufacturer of mains powered equipment may define power quality as those characteristics of the power supply that enable the equipment to work properly. These characteristics can be very different for different criteria.

Power quality is ultimately a consumer-driven issue, and the end user's point of reference takes precedence. Therefore, the following definition of a power quality problem is used in this textbook: any power problem manifested through deviations of voltage, current, or frequency that results in failure or misoperation of customer equipment.

Power quality, like quality of other goods and services, is difficult to quantify. Here is no single accepted definition of quality of electricity. There are standards for voltage and other technical criteria that may be measured, but the ultimate measure of power quality is determined by the productivity of end-user equipment. If the electric power is inadequate for those needs, this means that the "quality" is lacking.

Energy companies and end users of electric power are becoming increasingly concerned about the quality of electric power. The term power quality has become one of the most often used words in the power industry since the late 1980s.

There are four major reasons for the increased concern:

1. The increasing emphasis on overall power system efficiency has resulted in continued growth in the application of devices such as adjustable-speed motor drives and shunt capacitor for power factor correction to reduce losses. This is resulting in increasing harmonic levels on power systems.
2. The equipment of consumers with microprocessor-based control devices and power electronic devices is more sensitive to power quality variations than equipment used in the past.
3. End users have an increased awareness of power quality issues. Utility customers are becoming better informed about such issues as interruptions, sags, and switching transients. As a rule, they demands from suppliers' improvement of the quality of delivered energy.
4. Many components are now interconnected in a network. Integration processes mean that the failure of any component has much more important consequences.

The common thread running through all these reasons for increased concern about the quality of electric power is the continued push for increasing productivity for all utility customers. Manufacturers want more productive and more efficient machinery. Energy companies encourage this effort because it helps their customers become more profitable and helps defer large investments in substations and generation by using more efficient load equipment. Interestingly, the equipment installed to increase the productivity is also often the equipment that suffers the most from common power disruptions. In addition, the equipment is sometimes the source of additional power quality problems. When entire processes are automated, the efficient operation of machines and their control becomes increasingly dependent on power quality. The engineers are now attempting to deal with these issues using a system approach rather than handling them as individual problems.

Some discoveries that have had an impact on electric power quality have been done during the last decades:

1. Throughout the world, many governments have revised their laws regulating electric utilities with the intent of achieving more cost-competitive sources of electric energy. Deregulation of utilities has complicated the power quality problem. In many countries, there is no longer tightly coordinated control of the power from generation through end-use load. While regulatory agencies can change the laws regarding the flow of money, the physical laws of power flow cannot be altered. In order to avoid deterioration of the quality of power supplied to customers, regulators must expand their thinking beyond traditional reliability indices and address the need for power quality reporting and incentives for the transmission and distribution companies.

2. During last decades interest increases in distributed generation (DG), that is, generation of power dispersed throughout the power system. There are a number of important power quality issues that must be addressed as part of the overall interconnection evaluation for DG [11].

3. Solving power quality problem facilitated the creation in recent decades of new types of high-power electronic devices. New electronic devices not only eliminate those problems, but also permit the development of equipment that improves power-transmission systems.
   Based on power electronic technology we can, for example, control the magnitude and sign of the reactive power and hence the power factor $\cos\varphi$; improve the harmonic composition of currents and voltages; stabilize at specified points of the line; and ensure symmetry of the load currents and voltages in three-phase systems. (More information can be found, for example, in [9].)

4. Globalization of industry has heightened awareness of deficiencies in power quality around the world. Companies building factories in new areas are suddenly faced with unanticipated problems with the electricity supply due to weaker systems or a different climate.

5. Key figures have been developed to help with evaluation of power quality in various aspects. Regulatory agencies have become involved in performance-based ratemaking, which addresses a particular aspect, reliability, which is associated with interruptions. Some customers have established contracts with energy companies for meeting a certain quality of power delivery.

As was mentioned above, the energy companies and customers often have very different points of view. While both tend to blame about two-thirds of the events on natural phenomena (e.g., lightning), customers, much more frequently than personnel of energy companies, think that energy companies is at fault. When there is a power problem with a piece of equipment, end users may be quick to complain to the energy company. However, the energy company records may indicate no abnormal events on the feed to the customer. It must be realized that there are many events resulting in end-user problems that never show up in the energy company statistics. Transient overvoltage is most often the cause of disruption of power supply of consumers.

In addition, there are also power quality problems that may be related to hardware, software, or control system malfunctions. Electronic components can degrade over time due to repeated transient voltages and eventually fail due to a relatively low magnitude event. Thus, it is sometimes difficult to associate a failure with a specific cause. It is becoming more common that designers of control software for microprocessor-based equipment have an incomplete knowledge of how power systems operate and do not anticipate all types of malfunction events. Thus, a device can misbehave because of a deficiency in the embedded software.

In response to this growing concern for power quality, energy companies have programs that help them respond to customer concerns. The philosophy of these programs ranges from reactive, where energy companies responds to customer complaints, to proactive, where the energy companies are involved in educating the customer and promoting services that can help develop solutions to power quality problems. Since power quality problems often involve interactions between the supply system and the customer facility and equipment, regulators should make sure that distribution companies have incentives to work with customers and help customers solve these problems.

It is not always *economical* to eliminate power quality variations on the supply side. In many cases, the optimal solution to a problem may be replacement of a particular piece of very sensitive equipment to an equipment less sensitive to electricity quality variations. The level of electricity quality required is that level which will result in proper operation of the equipment at a particular facility.

Any significant deviation in the waveform, magnitude, frequency, or purity is a potential power quality problem. Of course, there is always a close relationship between voltage and current in any practical power system. Although the generators may provide a near-perfect sine-wave voltage, the current passing through the impedance of the system can cause a variety of disturbances to the voltage.

The ultimate reason that we are interested in power quality is economic value. There are economic impacts on energy companies, their customers, and suppliers of load equipment. The quality of power can have a direct economic impact on many industrial consumers. There has recently been a great emphasis on revitalizing industry with more automation and equipment that is more modern. This usually means electronically controlled, energy-efficient equipment that is often much more sensitive to deviations in the supply voltage than were its electromechanical predecessors. Thus, industrial customers are now more acutely aware of minor disturbances in the power system. There is big money associated with these disturbances.

The electric utility is concerned about power quality issues as well.

Meeting customer expectations and maintaining customer confidence are strong motivators for energy companies. With today's movement toward deregulation and competition between energy companies, they are more important than ever. The loss of a disgruntled customer to a competing power supplier can have a very significant impact financially on an energy company. Besides the obvious financial impacts on both utilities and industrial customers, there are numerous indirect and intangible costs associated with power quality problems. Residential customers typically do not suffer direct financial loss or the inability to earn income because of most power quality problems, but they can be a potent force when they perceive that the energy company is providing poor service. Home computer usage has increased considerably in the last few years and more transactions are being done over the Internet. Users become more sensitive to interruptions when they are reliant on this technology. The great number of complaints require energy companies to provide staffing to handle them. In addition, public interest groups frequently intervene with public service commissions, requiring the energy companies to expend financial resources on lawyers, consultants, studies, and the like to counter the intervention.

Many manufacturers of energy consuming equipment are also unaware of the types of disturbances that can occur on power systems. The primary responsibility for correcting inadequacies in load equipment ultimately lies with the end user who must purchase and operate it. Specifications must include power performance criteria. The useful service that energy companies can provide is dissemination of information on power quality and the requirements of load equipment properly operate [12].

6.3.4   Electromagnetic Compatibility

The problem of electromagnetic compatibility (EMC) is closely related with problem of power quality; very often, they are being considered together.

For example, in Russia power quality standards established by all-Union State Standard (GOST 13109-97 "Power quality limits in the supply systems"), are the levels of electromagnetic compatibility of power supply systems and electric grids of electricity consumers. *Under electromagnetic compatibility the ability of devices and equipment of energy consumers and all elements of electrical networks is understood to function under conditions of exposure to electric and magnetic fields and do not cause harmful interference to each other and to other objects.*

In other words, EMC is achieved when equipment and systems operate satisfactorily in the presence of electromagnetic disturbances. For example, the electrical noise generated by motor-driven household appliances, if not properly controlled, can cause interference to domestic radio and TV broadcast reception. The EMC affects the quality of life, operational reliability, and safety of life, where safety-related systems are present.

The electromagnetic environment in which a system operates may comprise a large number of different disturbance types, emanating from a wide range of sources including:

- mains transients due to switching,
- radio frequency fields of radio transmitters,
- electrostatic discharges from human body charging,
- surges, dips, and interruption of power supply in electrical network,
- magnetic fields from electrical networks and power—frequency transformers.

In addition to adequate immunity to all these disturbances, equipment and systems should not adversely add electromagnetic energy to the environment above the level of interference-free radio communication and reception.

Among *sources* are electromechanical switches, commutator motors, high-power semiconductor devices, digital logic circuits, and radio frequency generators.

The electromagnetic disturbances can propagate to the receivers through this path, in particular, this is case for a radio receiver, which contains a semiconductor device that responds to the disturbance and causes unwanted response, i.e., interference.

For many equipment and systems, EMC requirements now are an integral part of the overall technical performance specification. All apparatus placed on the market or taken into service must be immune to electromagnetic presented in the environment and must generate its own disturbance no greater than that permitting interference-free radio communication.

The EMC Directives currently in force in all developed countries refers to relevant standards which themselves define the appropriate immunity levels and emission limits. *The path* by which electromagnetic disturbance propagates from source to receiver comprises one or more of the following, Fig. 6.16:

Fig. 6.16 Coupling mechanisms for electromagnetic disturbance

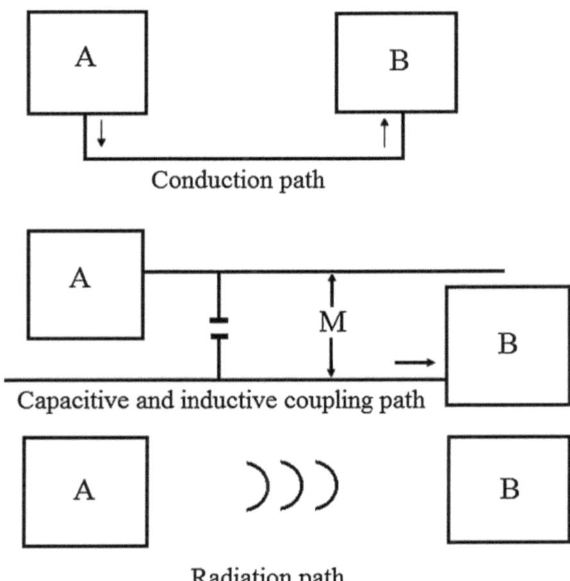

- conduction,
- capacitive or inductive coupling,
- radiation.

Conduction coupling is observed where there is a galvanic link between the two circuits, and dominates at low frequencies where the conductor impedances are low.

Capacitance and *inductive coupling* takes place between reasonably long colocation parallel cable runs. *Radiation* dominates where conductor dimensions are comparable with a wavelength at the frequency of interest, and efficient radiation is observed.

Designers and installers of electrical and electronic equipment need to be aware that all three coupling methods exist so that the equipment can be properly configured for compatibility. Analogue circuits respond adversely to unwanted signals in the order of millivolts. For malfunction of digital circuits, only a few hundred millivolts is usually sufficient. When high levels of transient disturbance in the order of several kilovolts are present in the environment good design is critical for compatibility.

To demonstrate how difficult is to solve the problem of electromagnetic compatibility of electric power source with its consumers, we now consider methods of protection against electromagnetic disturbances of one of the most sensitive devices —Printed Circuit Board (PCB.)

The field strength is higher for shorter distances and higher frequencies (shorter wavelengths).

There are four main methods for ensuring satisfactory EMC:

- shielding,
- cable screes terminations,
- PCB design and layout,
- grounding.

The preferred and most cost-effective approach to the achievement of EMC is to incorporate control measures into the design. At the beginning of the design stage, attention should be given to the basic principles of the EMC control to be applied in the design and construction of the product. The overall EMC design parameters can be determined directly or from the EMC standards, either as part of the delivery specification or as part of the legal requirements for market entry.

Two options for EMC control—***shielding*** and ***filtering*** and ***board level control*** are shown in Fig. 6.17a, b.

For shielding and filter solution, all external cables are or unscreened leads connected via a filter. The basic principle is to provide a clear—cut barrier between the inner surface of the shield facing the emissions from the PCBs and the outer surface of the shield interacting with the external environment. This solution requires utilization of a metal or metal-coated enclosure. If this is not possible, the board level control alternative is appropriate, where the PCB design and layout provide inherent barriers to electromagnetic energy transfer. Filters are required at all the cable interfaces, except where an effective screened interface can be used.

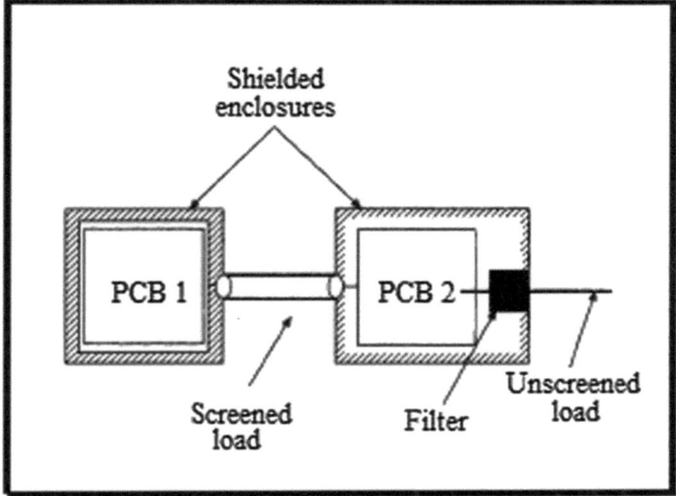

(a) Shield and filter solution

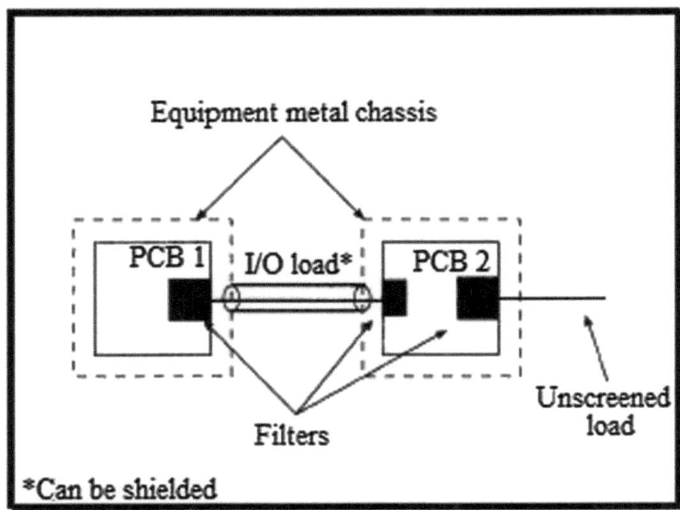

(b) Board level control solution

Fig. 6.17 Options for EMC control (*PCB* here is printed circuit board))

Emissions from the PCB tracks can be reduced at high frequencies if the devices have slow switching rates and slow transition (i.e. long rise and fall) times.

Choice of the device can also improve immunity by means of bandwidth control. The smaller the bandwidth, the less likely it is that high-frequency disturbances will appear within the ***passband of the circuit.*** Although rectification of the disturbance can proceed in the out-of-band region, the conversion process is more inefficient and higher immunity usually results.

The tracks on PCBs can act as antennas and the control methods should reduce their efficiency. The following methods can be used to good effect:

- reduce the area of all track loops,
- minimize the length of all high-frequency signal paths,
- terminate lines in resistors equal to the characteristic impedance,
- ensure that the signal return track is adjacent to the signal track,
- remove the minimum amount of copper on the board, i.e. maximize the surface area of the 0 V (zero-volt ground) and VCC (Voltage at the Common Collector) planes.

The latter two items are generally used where a multilayer board configuration is employed. These measures are highly effective; they reduce board emissions and improve circuit immunity to external disturbances. Multilayer boards are sometimes considered relatively expensive, but the extra cost is compared with the total costs of other measures that may be required with single or double-sided boards, such as shielding, filtering, and additional development and production costs.

Grounding is the method whereby signal returns are controlled, and it should not be confused with earthling aimed at protection from electrical hazards.

Grounding is important at both the PCB level and circuit interconnection level. The three main schemes of grounding are:

- series ground,
- single-point ground,
- multi-point ground.

The ideal scheme is the multipoint ground. Generally, single-point grounding is used to separate digital, analogue and power circuits. Multipoint grounding is used whenever possible within each category of analogue or digital circuit. Some series ground techniques may be employed where the coupled noise levels can be admitted.

Typically, the overall optimal solution is obtained from good basic design and successive experimentation.

Test methods for electromagnetic compatibility of equipment are regulated by international or national standards. For example, the EU countries used three types of EMC standards:

- basic standards,
- product-specific standards,
- generic standards.

Basic standards comprise the test methods and test levels at limits, but do not specify a product type.

Product standards comprise comprehensive details on how the product should be configured and operated during the test and what parameters should be controlled.

Generic standards are used in the absence of a product standard and concern all products, which may be operated within a defined environment. Both product and generic standards refer to basic standards for their test methods [13].

The problem of electromagnetic compatibility in modern world is so complex and actual that scientists in all developed countries deal with it, publishing a large number of original articles, reviews, monographs and textbooks. The state experts regulate the levels of electromagnetic interference suggest and measures to limit their impact. For example, in Russia more than 15 State Standards (GOST) address this problem. Comprehensive information on the problem of electromagnetic compatibility may be found, for example, in [14].

6.4 A Look at the Future Development of Power Transmission Systems

Below we consider three main areas for further development of the transport system of electrical energy: (1) development of micro-grids, intended primarily to serve the independent sources and independent consumers of electric power, (2) improvement or radical transformation of the powerful lines of a three-phase AC transmission, (3) expanding the scope and applications of DC transmission in the "big" power industry. Scientists and experts continue to search for radically new ways to transfer large flows of energy, but they have not yet come to power, which are required in the "big" power industry. One can talk about the laboratory stage of their development and application (microwave transmission, transmission using the fluxes of charged particles on the vacuum-processed paths, etc.).

6.4.1 Micro-grids

An increased portion of distributed generation stimulates creation of micro-grids (MG) often—called virtual power stations since they are, in fact, program packages for control over electricity demand and the distributed energy sources that allow the operator to model them as electricity resources. The MG allow the electric power companies to control over considerable number of consumers with large volume of energy consumption (energy capacity), influencing a set of their options concerning commercial operations. In this respect, the application of the MG provides a closer interaction of retail markets via the control over the system of the main transmission lines and the distribution system that form a bilateral electricity and money streams that provides deeply integrated system of optimization of everything that is necessary for efficient control over complex Smart Grids.

For consumers making decisions concerning the use of services of power supplying organizations and guided by the efficiency and utility criteria, creation of MG is a natural stage of creating conditions for the development of their own generating and accumulating capacities, first, non-polluting energy sources based on nontraditional renewable energy sources. The development of the MG, integrated both into a grid and the electric power and capacity market, will promote the increased role of the consumer in control over the power supply system.

By analogy with the central grid, the MG can generate, distribute, and regulate the electricity to the consumers. *Intellectual* MG include local sources of backup power supply and energy accumulation, are more flexible, and allow wider range of generating energy sources to be connected, including electric power stations harnessing wind, solar, and other energy types using URPS those integration represents a problem for the central electric power system.

The MG will be integral parts of the national electric power system: they are connected with regional grids and through them with the national electric grid. Electric power generation by small-distributed generators becomes economically expedient if there is a system *support* in the form of the main connection line. The electric power from micro-grids will be delivered to consumers and back into the regional grid depending on the supply and demand conditions. Real time monitoring and regulation will provide information exchange and will allow all supplies to be fulfilled on the national level. In this case, consumers can correct the electricity supply according to their requirements. Energy consuming devices inside residential buildings and factories are connected with a micro grid via systems of gauges and regulators.

In the last few years, great attention has been given to the development of micro grids of four basic types (designations):

(1) Remote MG isolated from the unified electric power grid and intended for power supply to remote consumers (settlement, military bases, or individual consumers);
(2) Micro grids of enterprises/campuses with individual owner;
(3) Commercial/industrial micro grids with several owners;
(4) Micro grids of municipalities and electric power supplying companies interconnected with larger microstructural objects.

Micro grids of the first type are, in essence, hybrid power complexes. Their distinctive feature (except the first type) is the ability to be disconnected from the grid belonging to a power supplying company in the case of voltage dip or failure.

By 2015, about 2000 MG with total capacity more than 3.1 GW have been created. In 2010, their number was less than 100 MG. In the process of MG exploitation, contradictions appeared between the electric power suppliers and consumers on problems of possession and control. Search is underway of ways of their elimination [15].

6.4.2   Strong Grid on the Basis of Flexible Alternative Current Transmission Systems

Fast growth of distributed electric power engineering, first of all renewable (8% annually), in the developed countries with energy shortage necessitated essential changes in electric power engineering and comprehension of advantages of integration of local power systems on the state and interstate levels based on unified administrative and information technology, i.e. on construction of Smart Grid.

Many developing countries with severely limited economic potential should choose from two variants or their combination:

- The first variant more simple for implementation envisages overcoming the technological gap by modernization of existing grids and their transformation into the Strong Grid with large throughput and increased reliability.
- The second variant, more difficult and expensive, envisages implementation of the Smart Grid concept providing complete automation of the process of electric power transmission and distribution and radical change of the principles of interaction of the grid and generating companies with consumers. In this case, the principle of succession and technological compatibility should be one of the key principles: the modernized equipment of electric power companies must be matched with new technologies and integrated into a new electric power system.

Among problems of transformation of existing (traditional) electric grids, the important place is occupied by high-power electronics, automatic control, and protection systems of new generation for solving problem of full observation and control of Unified Energy System (UES) electric regimes in real time. This should essentially increase controllability and efficiency of UES and provide an increase in the reliability of electrical supply of consumers to 0.9990. Wide introduction of Flexible Alternative Current Transmission System (FACTS) and perfection of complexes for automatic emergency protection and operator control is provided. The FACTS comprises alternating current transmission systems equipped with modern high-power electronic devices. The FACTS technology transforms the function of the electric grid from the existing *passive* into the *active* one.

The technical FACTS base (devices) is formed by:

- Longitudinal compensating devices (conventional capacitors and capacitors regulated by thyristor groups,
- Static thyristor compensators,
- Direct current inserts,
- Electromechanical frequency converters based on a synchronous machines,
- Controllable shunt reactors,
- Synchronous compensators,
- Energy storage devices.

Functions of some of them are given in Table 6.4.

Table 6.4 Controllable grid parameters for indicated FACTS devices

FACTS device	Controllable parameters
1	2
Static synchronous compensator (STATCOM without energy storage device)	Power control, reactive power compensation, smoothing of fluctuations, and maintenance of constant voltage
Static synchronous compensator (STATCOM with electric energy storage device (BESS, SMES, and large direct current capacitor)	Voltage control, reactive power compensation, smoothing of fluctuations, increase of static and dynamic stability components, maintenance of constant voltage, and AGC
Static compensator for reactive power (SVC, TCR, TCS, and TRS)	Voltage control, reactive power compensation, smoothing of fluctuations, increase of static and dynamic stability components, and maintenance of constant voltage
Thyristor-controlled braking resistor	Smoothing of fluctuations, and increase of static and dynamic stability components
Static synchronous longitudinal compensator (without electric energy storage device)	Control of currents, smoothing of fluctuations, increase of static and dynamic stability components, and limiting short-circuit currents
Static synchronous longitudinal compensator (with a battery)	Control of currents, smoothing of fluctuations, and increase of static and dynamic stability
Thyristor controlled series capacitor	Control of currents, smoothing of fluctuations, increase of the static and dynamic stability, and limitation of short-circuit currents
Thyristor-controlled series reactor	Control of currents, smoothing of fluctuations, increase of static and dynamic stability, maintenance of a constant voltage, and limitation of short-circuit currents
Thyristor controlled phase shifting transformer	Control of power overflows, smoothing of oscillations, increase of the static and dynamic stability, and maintenance of constant voltage
Unified controller of power overflows	Control of active and reactive power overflows, voltage control, compensation for reactive power, smoothing of fluctuations, increase of static and dynamic stability, maintenance of constant voltage, and limitation of short-circuit currents
Thyristor-controlled overvoltage limiter	Transient overvoltage protection

(continued)

Table 6.4 (continued)

FACTS device	Controllable parameters
Thyristor-controlled voltage regulator	Control of reactive power overflows, voltage control, smoothing of fluctuations, increase of static and dynamic stability, and maintenance of constant voltage
Interline controller of power overflows	Control of reactive power overflows, voltage control, smoothing of fluctuations, increase of static and dynamic stability, and maintenance of constant voltage

Nowadays the FACTS devices, as a rule, are taken to mean a set of devices of an electric grid intended for stabilization of voltage, increase of controllability, optimization of distribution of energy flow, decrease in losses, damping of low-frequency fluctuations, increase of static and dynamic stability, and as a result, increase in the grid throughput and decrease in losses. The essential role in all variety of FACTS devices is played by high-power electronics based on various modifications of voltage converters using controllable semiconductor valves. Wide introduction of easily adjusted or self-adjusted innovative elements of high-power electronics and new generation of converter equipment, the latest technologies in the field of high-temperature superconductivity (cables and storage devices), and microprocessor systems of automatic control and regulation (yet on the limited scales) allow new qualities to be given to already existing grids.

The Strong Grid concept based on the FACTS technology possesses important advantages: (a) introduction of modern equipment and means in already existing grids allows large expenses for construction of lines with high throughput to be avoided, (b) the technology admits (and even envisages) gradual (step-by-step) expansion of allowable power limits of transmission lines in the process of investments where required and when necessary. It allows to provide in advance the progressive scenario of sharing of mechanical switchboards and gradually entered controllers FACTS so that to reach an object in view by stage-by-stage investment. In the conditions of financial restrictions, it is a solid argument in favor of Strong Grid. Gradual innovative technical and technological transformation of EPS, and corresponding updating of basic principles and the purposes and problems of development of electric power engineering will transform Strong Grid in Smart Grid.

6.4.3   Smart Grid

According to the concept of *Smart Grid*, the future power system is considered as an infrastructure similar to the Internet intended for maintenance of power, information, economic and financial relations between all participants of the power market and other interested parties. Smart Grid is the concept of innovative transformation of electric power engineering as a whole rather than of its individual functional or technological segments based on revision of the existing basic principles, purposes, problems in the development of electric power engineering and scales following from it, and character of problems to be solved considering predicted social, economic, scientific, technical, ecological, and other consequences of their implementation [4, 16, 17].

For example, in the USA such program has the national status and is implemented with direct support of the country leaders. In the EU countries, the Smart Grid Technological Platform European Power System of the Future was organized still in 2004 for coordination of works and elaboration of uniform strategy of the development of electric power engineering. Its ultimate goal is elaboration and implementation of the program for developing the European Power System until 2020 and later.

The main ideologists of working out of the concept were the USA and the EU countries; it was subsequently recognized and developed practically in all large industrially developed and dynamically developing countries that invest considerable money for its implementation: China—70, the USA—19, India—10, the EU countries—7.0, Great Britain—3.0, Austria—1.0, Canada—0.5, and South Korea —0.3 billion of US dollars.

Smart Grid represents an integral automated mechanism uniting manufacturers of the electric power, electric grids, and consumers. This mechanism is controlled by a computer center which collects data on electric power consumption level from millions digital controllers in real time. Specialized software helps to trace the operating regime of all participants of the process of electric energy generation, transmission, and consumption. Table 6.5 shows the basic connection technologies according to Smart Grid concept.

The main advantage of the Smart Grid is that it automatically responds to changes in various parameters in the power supply system and allows uninterrupted electrical supply with maximum economic efficiency. In this case, the influence of the human factor in the Smart Grid operation is reduced to a minimum.

Table 6.5 Integrated connections

Technology type	Basic components
Wireless technologies	• Multiaddress radio system • Notification grids • Radio systems of expanded spectrum • Wi-Fi • WiMAX • Cellular structure of next generation • Time partition multiple access • Code division multiple assess (CDMA) • Small satellite terminal
Other technologies	• Internet of new generation (Internet-2) • Broadband power line (BPL) • Grid with optical cable delivered to the user • Fiber-optical coaxial cable • Radio frequency identification (RFID)

In fact, the Smart Grid is a combination of possibilities of information technologies that already became habitual in many spheres of industrial activity with high-power electronics and electrical power engineering.

Within the limits of the Smart Grid concept a variety of requirements of all interested parties (state, consumers, regulators, power companies, sales and municipal institutions, proprietors, manufacturers of equipment, etc.) is reduced to a group of the so-called key requirements (values) of the new electric power engineering formulated as follows:

Availability—supply of consumers with energy without restrictions depending on when and where it is necessary for them and on its paid amount;

Reliability—possibility of overcoming negative physical and information consequences without total power switching-off or high expenses for operation recovery, restoration (self-restoration) as fast as possible;

Profitability—optimization of tariffs for electric energy for consumers;

Efficiency—maximization of the efficiency of utilization of resources of all kinds, technologies, and equipment in the process of power generation, transmission, distribution, and consumption;

Friendliness to the environment—the maximum possible decrease of negative ecological effect;

Safety—avoiding situations in electric power engineering dangerous to people and environment.

Table 6.6 compares the properties of the modern EPS and the system based on the Smart Grid concept.

Table 6.6 Properties of the modern and perspective electric power systems

Present-time electric power system	Electric power system based on the Smart Grid concept
Unilateral connection between elements or its absence	Bilateral connections
Centralized generation—distributed generation difficult to integrate	Distributed generation
Predominantly radial topology	Predominantly grid topology
Response to the consequences of an accident	Response to accident prevention
Operation of the equipment before failure	Monitoring and self-diagnostics extending the operating lifetime of the equipment
Manual recovery	Automatic recovery—self-repairing grids
Propensity to accidents of the system	Prevention of system accidents
Manual and fixed disconnection of the grid	Adaptive disconnection
Testing of the equipment in place	Remote monitoring of the equipment
Limited control over the power	Control over the power redistribution
Inaccessible or much belated information on the price for the consumer	Price in real time

The Smart Grid concept, leaning against the strategic vision of electric power engineering in future, involves principles of construction of such grids and key requirements to them from which functional properties (characteristic) also follow: administrative, technological, standard, and informational.

Simplified diagram of the implementation of the concept Smart Grids is shown in Fig. 6.18.

As already indicated above, only grids do not limit the Smart Grid concept—it covers all links of the technological chain from electric power generation to its consumption, Fig. 6.18.

It provides for each of them the achievement of the following purposes using the corresponding means:

Generation—increase of the reliability and profitability of electric power production using modern highly intelligent control and management devices, including

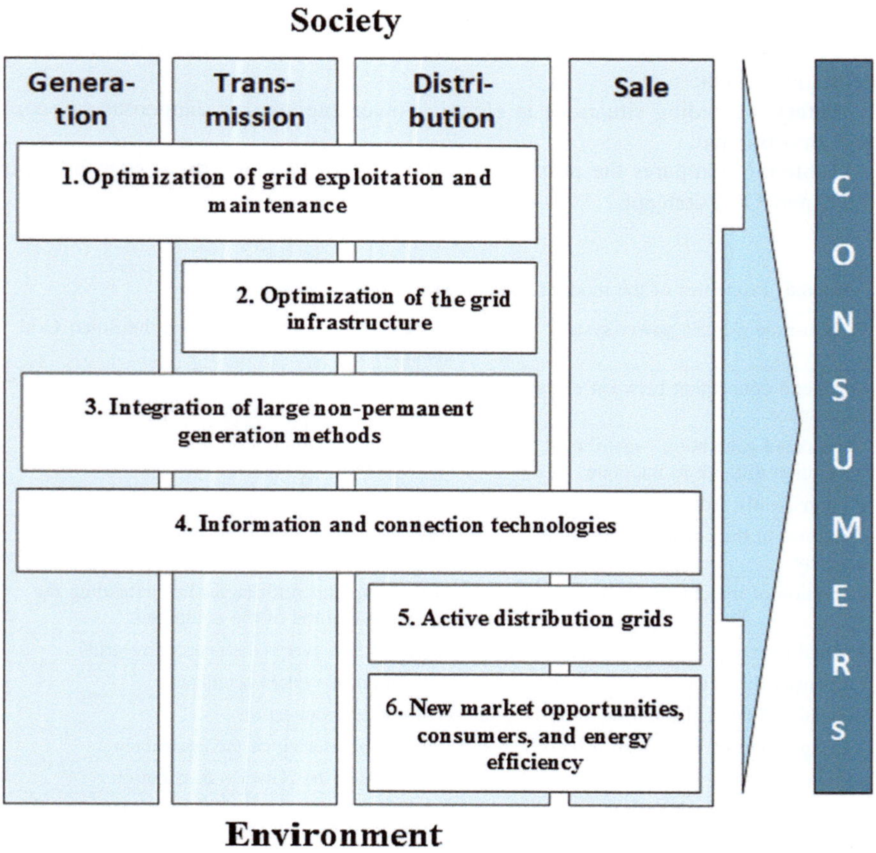

Fig. 6.18 Priorities in the development of smart grids in the EU countries

IT, integration of renewable energy sources, distributed generation and energy storage devices;

Transmission electric grid—provision of reliability of electric power transmission and electric grid controllability via large-scale monitoring of regimes and control over them taking advantage of new means and technologies (FACTS, Power Management Units (PMU), artificial intelligence, etc.) as well as via increased use of pilotless flying machines for control over technical PTL state as scheduled (every 1.5 years) and off-scheduled (after each natural anomaly) inspections;

Substations—maintenance of the reliability and controllability of substations at the expense of modern electro technical equipment and automation based on modern diagnostic tools, monitoring, and management using information and computer technologies;

Distributive electric grid—increase of controllability and reliability by means of application of distributed automatics and protection systems on modern microprocessor basis using new information, computer, and Internet technologies;

Consumers—their supply with high-intellectual systems for monitoring and accounting of the consumed electric power, regulation of power consumption and management by loading, including emergencies.

Creation of an *intellectual grid* provides application of a large set of new technical means and processing methods (Their list and characteristics can be found in the special literature).

Attempt to provide a quantitative estimation of the effect from implementation of the Smart Grid concept as of 2025 be undertaken in the USA, Table 6.7.

As of the end of 2010, there were 90 pilot projects of Smart Grid creation all over the world. In the process of their implementation, it becomes obvious that the Smart Grid concept carried out on the national stage undergoes considerable changes caused by distinctions in regulation regimes, available infrastructure of electric power systems, and national economic priorities. In the countries with limited mineral and power resources, the strategy is largely focused on the creation of favorable conditions for the development of renewable power sources, stimulation of energy saving, and increase in the efficiency of consumption of power resources.

For Russia with its large reserve of power resources, huge length of electric grids, high degree of depreciation of the equipment, the problem of maintenance of reliability and efficiency of the electric grid complex operation with limited investments and deficiency of time occupies the first place. The concept itself envisages the creation of power supply systems with the combined grid—combination of Strong Grid and Smart Grid.

The scale of expected transformations in the process of future intellectualization of electric power engineering is comparable with that of revolutionary changes in communication and information spheres that have made a habitual reality the Internet, mobile communication, and a number of other contemporary achievements that unrecognizability changed everyday life.

Table 6.7 Effects of implementation of the Smart Grid concept

Parameters	2000	2025		
	Basis	Electric power system without Smart Grid (scenario 1)	Smart Grid based electric power system (scenario 2)	Ratio of the parameters of scenario 2 to those of scenario 1 (%)
Electric energy consumption (blrd kW h)	3.800	5.800	4.900–5.200	10–15, decrease
Energy consumption per unit GNP (kW h/dollar GNP)	0.41	0.28	0.20	29, decrease
Decrease in demand in peak loading (%)	6	15	25	66, increase
Emission of CO_2 (billion tons of carbon)	590	900	720	20, decrease
Level of productivity growth (%/year)	2.9	2.5	3.2	28, increase
Actual GNP (blrd dollars)	9.200	20.700	24.300	17, increase
Business economic loss (blrd dollars)	100	200	20	90, decrease

6.4.4   Direct Current Transmission Lines

It is interesting to note that in the period of fast development of three-phase high-voltage (up to 150 kW) transmission lines, M.O. Dolivo-Dobrovolskii, based on technical-economic calculations, concluded that when power was transmitted at a distance of a few hundred kilometers at voltage above 200 kV, it would be expedient to generate and to distribute the alternating current and to transmit the high-voltage direct current. The DC line at the beginning and end should be connected to an inverter substations equipped with mercury rectifiers. He had come to this conclusion not knowing at all such problem of high-power AC transmission lines as stability.

Nowadays his prediction is justified, and in many countries, super-high-voltage DC OTL lines operate successfully, Fig. 6.19. Although the bulk of the world's electric transmission is carried on AC systems, recent progress in high-voltage direct current technology has enabled the development of large-scale DC transmission by overhead lines and submarine cables, which have become economically attractive in long-distance transmission of large bulk of power. Currently, numerous installations with voltages up to 800-kV DC are in operation around the world.

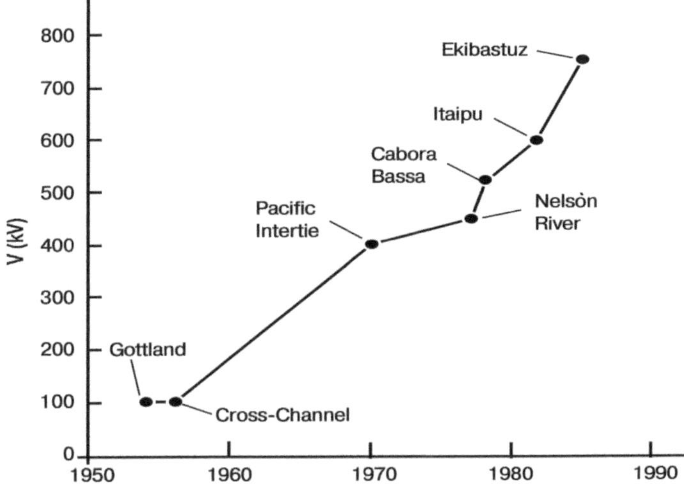

Fig. 6.19 Major DC transmission system in chronological order of their installations

In 1954, the Swedish State Power Board energized the 60-mile, 100-kV DC submarine cable utilizing U. Lamm's mercury arc valves at the sending and receiving ends of the world's first high-voltage direct current (HVDC) link connecting the Baltic island of Gotland and the Swedish mainland. Figure 6.20 shows the external appearance of the experimental DC OTL of the highest class of 1500 kV, and in Fig. 6.21—the appearance of the operating DCOTL.

Currently, numerous installations with voltages up to 800 kV dc are in operation around the world. Only in one decade (from 1987 to 1997), about 50 DC high-voltage lines went into operation around world. The power of many DC high-voltage lines is as much as 2000 MW, at voltages of 600 kV and above. By 2010, the total power transmission DC approached 40 GW. In Europe, to date, 24 DC OTL have been built and inserted with total capacity of 12.5 GW. Note that, in DC-transmission, the power losses in the line are 20–30 times less than those in AC-transmission. That is why the issue of energy losses in DC transmission lines in Sect. 6.3.2 is not considered.

When using DC-links, one option is to introduce an array of voltage inverters on the DC-side. A bank of capacitive batteries is attached to the DC-buses. As a result, the actual DC-transmission line disappears. This short link is often known as a B2B (back-to-back) system. Thus, in this case, there is practically no transmission line due to localization of the converters. The routs of the DC cable lines in Europe are shown in Fig. 6.22.

The development of dc power transmission systems is constrained by a number of shortcomings specific to them:

- high price, limited overload capacity converters (rectifiers and inverters), and switches (For example, circuit breakers HVDC are difficult to manufacture, since they require the presence of a built-in switch mechanism to reset the current, or the arc will be formed and are accelerated wear contacts);

Fig. 6.20 Photograph of DC OTL-1500 kV (experienced plot, Russia)

- high price of high-voltage cables with a rather complex design necessary to overcome the large water barriers (oceans, seas, and large lakes), Fig. 6.23.
- complexity of creating a control circuit used for switching on and off thyristors of HVDC line; with small length of the HVDC line the losses in the converters can be greater than in the AC power lines;
- Problems with the implementation of multi-terminal systems. Management of power flow in a multi-terminal DC system requires good switching devices between all terminals.

As seen in Fig. 6.24, the cost of UHV DC line when the transmission range is 500–1200 km is close to the value of the UHV AC line. At a distance of about 2000 km, the cost of UHV DC line ± 600 kV transmission capacity of 4000 MW is already lower than the cost of UHV AC line by 1.4 times. The length of UHV DC

Fig. 6.21 Photograph of DC OTL in North Dakota (USA)

line and UHV AC of equal value for the line 4000 MW ultra-high voltage is about 900 km and of the lines of 1000 MW—around 600 km.

Electric power transmission system using a direct current is becoming somewhat simpler and less expensive when using short link often known as a B2B (back-to-back) system. When using DC-links, one option is to introduce an array of voltage inverters on the DC-side. A bank of capacitive batteries is attached to the DC-buses. As a result, the actual DC transmission line disappears. Thus, in this case, there is practically no transmission line due to localization of the converters.

In future, the pace of development of DC cable systems of power transmission over long distances will largely depend on the success in the development of industrial-scale high-temperature superconductivity (HTS) [18]. The complexity and high cost of providing the liquid helium environment prevented commercial use of the phenomenon of superconductivity at liquid-helium temperatures, which was discovered in 1911.

The situation began to deteriorate drastically since the late nineties of the XXth century. In late 1986, a ceramic material LaBaCuO was discovered to be superconducting at 35 K and in 1987; the material YBaCuO was found to have a critical temperature of 92 K. Since that time, the critical temperatures of these HTS materials have progressively increased to over 130 K. The enormous significance of these discoveries is that these materials will be superconducting in liquid nitrogen, which has a boiling point of 77 K and is much easier and cheaper to provide than helium.

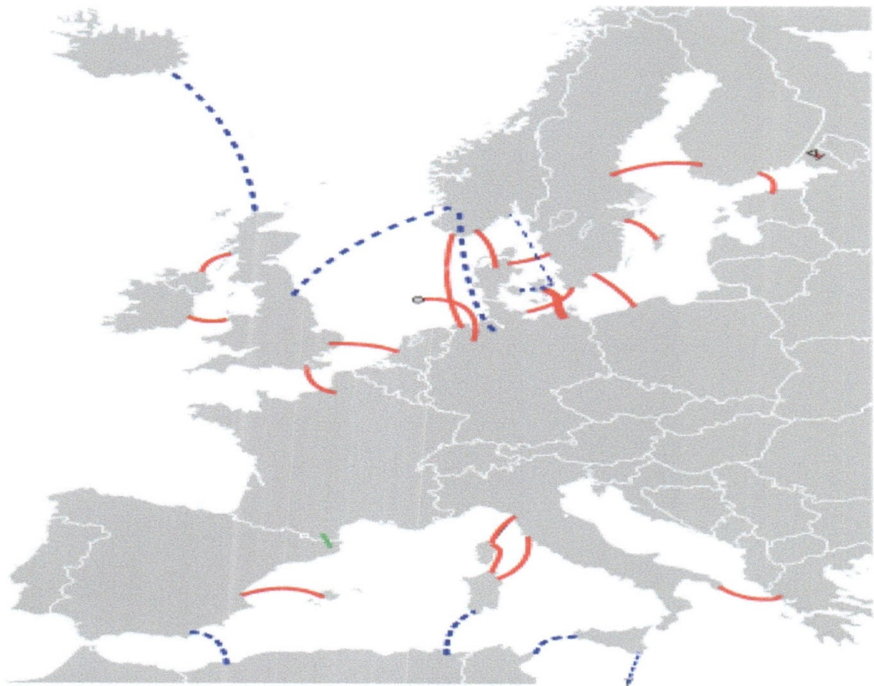

Fig. 6.22 DC cable lines in Europe (*Red* marked existing lines, *green* under construction, *blue* the proposed/discussed)

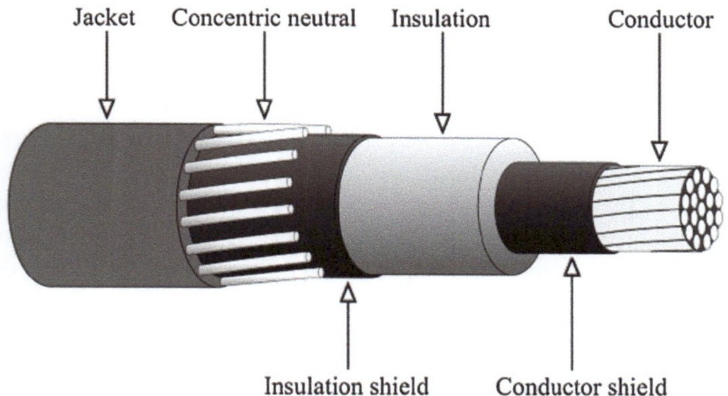

Fig. 6.23 A concentric neutral cable, typically used for underground residential power delivery

Much work has been directed towards finding materials with higher T_c values but this has remained at 133 K for some time. However, considerable effort with resulting success has been directed to the production of suitable HTS conductors.

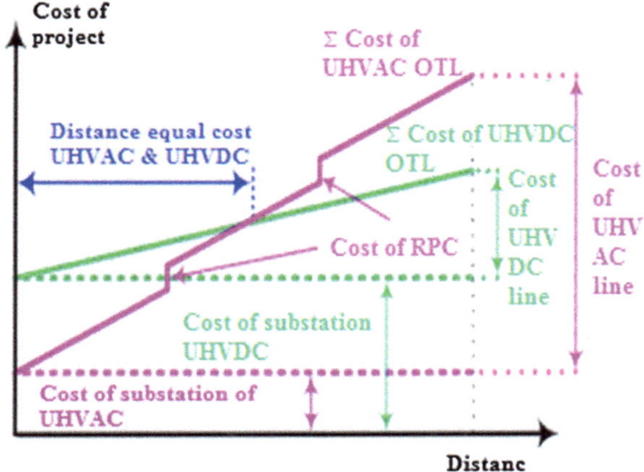

Fig. 6.24 Comparison UHV AC with UHV DC transmission lines by their costs

The HTS material is very brittle and it is deposited using laser deposition onto a suitable substrate tape.

The tape is 3 mm wide and cables of up to 600 m in length have been produced.

There are many trials being made of the application of the HTS cables throughout the world including USA, Europe, and Japan. There are prototypes of underground power cables, power transformers, large motors, and generators, inductive energy storages (see Sect. 7.2.3) and others in active development and in use.

Detroit is being re-equipped with HTS power cable for its transmission system and copper cables weighing over 7 tons are being replaced with HTS cables of less than 0.12 tons. The electricity supply of the city Geneva in Switzerland is completely provided by power transformers wound with HTS conductors. It is expected that there will be definite power saving from the use of HTS.

In conclusion to Chap. 6, it must be emphasized that today great concern of power engineers all over the world is not the transportation of electricity, but provision by primary energy resources and the development of energy efficient and environmentally safe methods of converting them into electricity and heat.

Questions and Tasks to Chapter 6

(1)   Where and when were the first power systems created?
(2)   What are the main steps of creating power systems?
(3)   Name the three largest power systems linking national systems of European countries.
(4)   What are the basic aims, which underlie the creation of the systems of power supply on continental and intercontinental scales?
(5)   What requirements must be met in order to ensure reliable operation of an electricity grid?
(6)   What is the difference between the specific energy losses in electric networks of different countries with various levels of technological development?

(7) Types and quantity (as a fraction of the total) of electricity losses during its transportation.

(8) Key measures to reduce losses during electricity transportation.

(9) What is commonly referred to by the term "power quality"?

(10) Why are we concerned about power quality?

(11) What are the main sources of electromagnetic disturbances?

(12) In which cases one or another form of EMD (electromagnetic disturbances) is dominated?

(13) Name the four main methods of protection from EMD and give them a brief description.

14) Types and purposes of microgrids.

(15) What is a "Strong Grid"?

(16) The basic concept of "Smart Grid": principles of construction and operation, the advantages in comparison with traditional power grid.

(17) What would you say about strong points and weaknesses of direct current transmission lines?

(18) Why the first attempts to apply on a large scale the direct current for electric energy transmission over long distance were unsuccessful?

(19) Does the ratio values of HVDC and HVAC lines vary with the length of the transmission line?

(20) What is the main function of the direct current cable transmission lines?

(21) What are basic aims, which underlie creation of the systems of power supply on continental and intercontinental scales?

References

1. Warne DF. Newness electrical power engineer's handbook. Amsterdam: House Elsevier; 2005.
2. Meier A. Electric power systems: a conceptual introduction. Hoboken, New Jersey: IEEE Press, Willey Nescience; 2006.
3. European Commission Directorate-General for Research Information and Communication Unit European Communities. European Technology Platform Smart Grids, Vision and Strategy for Europe's Electricity Networks of the future. European Communities; 2006.
4. Grigsby LL. Power system stability and control. Boca Raton, London, New York; CRC, Taylor and Francis Group; 2007.
5. Ushakov VY, Kharlov NN, Chubik PS. Energy saving in the enterprises of fuel and energy complex: textbook (ed. V. Ushakov). Tomsk: Tomsk Polytechnic University, TPU Publishing House; 2017.
6. Grigsby LL (ed.). Electric power engineering handbook, 2nd ed. RCC Press; 2006.
7. Hunt S, Shutteworth G. Competition and choice in electricity. Chichester, England: Wiley; 1996.
8. Paul C. Electromagnetic compatibility. USA: Wiley; 1988. p. 426.
9. Ben Elghali SE, Benbouzid MEH, Charpentier JF. Marine tidal current electric power generation technology: state of the art and current status. In: IEEE international conference on electric machines and drives 2007, IEMDC'07. 2007. vol. 2, p. 1407–12.

10. Dugan RC, Mc Granaghan MF, Santoso S, Reaty HW. Electrical power systems quality, 2nd ed. 2004. World development indicators 07. Washington, DC: The World Bank; 2007.
11. El-Hawary ME. Electrical energy systems. CRC Press; 2000. p. 365.
12. Termuchlen H, Empsperger W. Clean and efficient coal fired power plants. New York: ASME Press; 2003.
13. Ott HW. Electromagnetic compatibility engineering. Livingston, NJ: Wiley; 2009. p. 843.
14. Engineering Guide for Integration of Distributed Storage and Generation, Prod. ID: 1024354, 2012—p. 176.
15. Bergeron R, Slimani K, Lamarche L, Cantin B. New architecture of the distribution system using electronic transformer. In: ESMO-2000, panel on distribution transformer, breakers, switches and arresters. 2000.
16. Bush SF. Smart Grid: communication-enabled intelligence for the electric power grid. Willy-IEEE Press; 2014.
17. Narlikar AB. High temperature superconductivity. Springer; 2004.
18. Ter-Gazarian A. Energy storage for power systems, 2nd ed. IET; 2011.

Chapter 7
Energy Accumulation (Store)

7.1 Systematization

At present, the electric power is the product that is compelled to be consumed during manufacture. The use of energy storages (ES) provides division in time of the processes of electric power generation and consumption and, as a consequence, [1]:

(1) Improves the maneuverability of the ES for load capacity and due to this allows:

- to reduce capital investments in generating capacities (by about 30%),
- to reduce exploitation expenses for basic capacities,
- to reduce losses from intersystem overflows of excessive capacities during periods of minimal loading;

(2) Improves the EPS technical and economic characteristics, helps to implement the Smart Grid concept thanks to the following:

- provision of static and dynamic stability and hence, increase in the reliability of EPS operation,
- maintenance of the voltage frequency and level and improvement of electric power quality,
- increase of the throughput of intersystem connections.

Among the ES parameters that define their suitability for EPS are:

- maximum capacity,
- total power consumption,
- operating time,
- time of power reverse (change of "charging–discharging" and "discharging–charging" regimes).

A typical scheme for connecting an electrical energy storage device to a grid is shown in Fig. 7.1.

© Springer International Publishing AG 2018
V.Y. Ushakov, *Electrical Power Engineering*, Green Energy and Technology,
https://doi.org/10.1007/978-3-319-62301-6_7

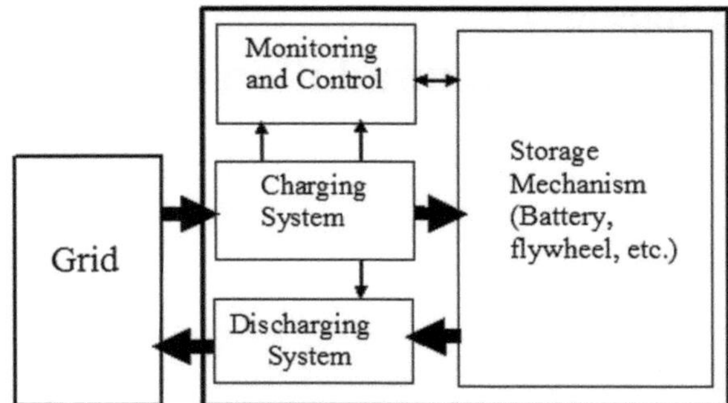

Fig. 7.1 Interaction of basic components of an energy storage system

Owing to a wide variety of ES, it is expedient to classify them according to the accumulated energy type:

1. *Electric energy* storages (EES) form a class of storages accumulating directly electric energy. They possess the highest rate and the widest range of power consumption. Thanks to these properties, they are suitable for increasing the stability and reliability of EPS operation and alignment of loading schedule. They are reliable in operation since they have no moving parts, are insensitive to the connection place either in the centre of the load or directly at the consumer. The EES involve:

    1.1. Capacitor storages (CSs),
    1.2. Electrochemical storages—storage batteries (SBs),
    1.3. Superconducting inductive energy storages (SIESs).

2. *Potential* and *kinetic energy* storages:

    2.1. Pumped storage power plants,
    2.2. Air-compressor energy storages,
    2.3. Inertial energy storages (flywheels and super-flywheels),
    2.4. Electromechanical energy storages.

7.2 Electric Energy Storages

7.2.1 Capacitive Energy Storages

Capacitors are simple devices: two metal plates sandwiched around an insulating dielectric. When charged to a given voltage, opposing charges fill the plates on either side of the dielectric. The strong attraction of the charges across the very short distance separating them makes a tank of energy. Capacitors oppose changes in voltage; it takes time to charge the plates, and once charged, it takes time to discharge the voltage.

The most widespread capacitor storages of electric energy are conventional radio engineering *capacitors*. However, as applied to electric power engineering, they have the following main disadvantages: low working voltage, low specific stored energy density and hence small (in comparison with other storage types) capacitance, and a minor of energy storage that seldom exceeds several hours. It took many years of research and development in the field of manufacture of capacitors to produce the capacitors that meet the requirements of electric power engineering in the greatest possible measure. The developers succeeded in increasing the specific capacitance of the capacitors by 10^8 times (several tens of farads per 1 cm^3) and the energy storage time by an order of magnitude (up to ~ 100 h). The modular design allows high-power capacitor batteries at several tens of kilovolts, several kilo amperes, and power capacity 10^{11}–10^{12} J to be developed. The block diagram of the CSs based on a capacitor battery is shown in Fig. 7.2.

At present, the field of applicability of the capacitor energy storages in electric power engineering is still limited by their disadvantages that have not yet been eliminated. The main disadvantages are the following:

- High price,
- Necessity of changing the polarity when proceeding from the charge to the discharge regime and *vice versa*,
- Devices of control over CS operation based on thyristor converters generate higher harmonics that distort sinusoidal waveform of alternating current.

As a result, the field of application of the capacitors in electric power engineering is limited mainly by short-term electric power accumulation and current rectification, correction, and filtration in electric power engineering circuits.

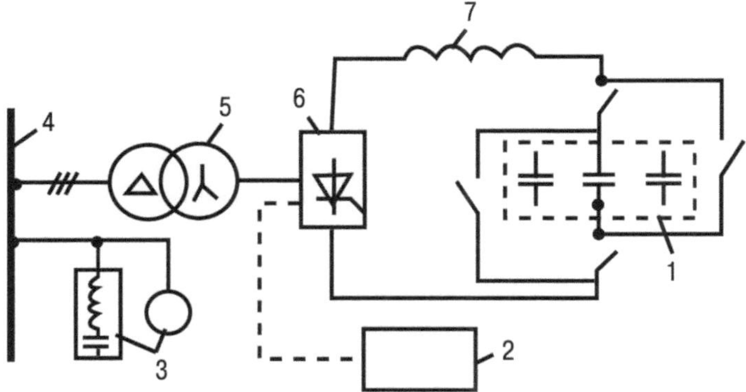

Fig. 7.2 Block diagram of the CS comprising capacitor battery *1*, control system *2*, smoothing reactor *3*, alternating current EPS *4*, transformer *5*, control unit *6*, compensating filter system *7*

7.2.2   Storage-Batteries

They use electric energy for implementation of chemical reactions that converts it into chemical energy in the charge regime and returns it to the same form of electric energy in the discharge regime. In this case, the composition of the electrode material is changed, and the electrolyte is spent. Such storages are called *accumulators*, and are combined in groups; they are called storage batteries or secondary battery systems.

Nowadays a number of companies have started production and sufficiently wide practical application of high-capacity storage batteries in electric power engineering.

Electrochemical storages have sufficiently high efficiency (65–70%), their capacity ranges from 100 mAh to 2000 Ah and specific energy capacity (200–300 kWh/m^3); the operating lifetime for perspective models is predicted on a level of 20 years. This will depend, among other things, upon the number of charge–discharge cycles and the type and construction of the battery employed.

Among the disadvantages of the electrochemical accumulators are a limited number of *charge-discharge* cycles (in most cases 1–2 thousands), sensitivity to temperature, long charging time sometimes exceeding by several orders of magnitude the discharge time, inadmissibility of deep discharge for lead accumulators and, on the contrary, necessity of the complete charge-discharge cycle for metal hydrate and many other accumulator types. The charge storage time is also rather limited—from several days to several months.

These disadvantages can be levelled in an electric power complex including the electrochemical energy storage and the ***electrochemical generator (ECG)***. The electrochemical generator is the power complex including systems of fuel preparation, conditioning, energy output, waste recycling, etc. The key element—the heart of such generator—is actually the energy converter—***fuel element*** (FE) representing a galvanic cell in which the electric energy is generated in the course of oxidation-reduction transformations of reagents arriving from the outside (it was patented in the first half of the XIXth century). It directly transforms the fuel energy into electricity, passing by ineffective, proceeding with high losses, burning processes (Biochemists have established that the biological hydrogen-oxygen fuel element is built into each living cell). The FE can be referred to chemical energy storages, since with its help the synthetic fuel generated by electric power under minimal loading (for example, hydrogen produced by electrolysis) is transformed into the electric power required under maximal loading. Nowadays this device is especially attractive for creation of effective energy source and hence is considered in Chap. 8 of this manual entitled "Unconventional methods of electric energy production."

7.2.3 *Superconductive Inductive Energy Storages*

Inductive energy accumulation is the most effective in superconducting magnets, since energy accumulation and output in them are practically not accompanied by losses. The principal advantage of the inductive storages consists in the fact that the energy in them is reserved in the same type in which it is used—electromagnetic. There is no need to convert one energy type into another, which provides high efficiency (97–98%) and fast response of the device. Therefore, the unique properties of the superconductive inductive energy storage (SCIES) are the capability of almost instantaneous transition from the energy accumulation regime to the regime of its delivery and a high speed of power takeoff. This property of the SCIES was first in demand in high power physics and pulsed power.

Discovery in 1957 of superconductors of the second type possessing high values of critical current density and magnetic induction caused avalanche increase in interest to the practical application of this phenomenon. Under the heading of conductors, there are continuing advances in superconductors, which are now able to operate at liquid nitrogen temperatures.

In all advanced countries (the USA, the USSR, Great Britain, France, Japan, and Germany), works were underway on creation of both individual elements of superconducting systems (wires, cables, and induction coils) and systems themselves (SCIES, superconducting magnets for accelerators, etc.). These works gave a powerful impetus to the perfection of accelerators, thermonuclear installations, cable lines, etc. Almost simultaneously, work was started in the interests of the electric power engineering.

High rate of SIES response in EPS allows active and reactive power to be controlled synchronously. The SIES can store energy as much as long in the form of magnetic field energy. This allows systems with high level of readiness (time from giving a command to energy delivery to a load is ~ 1 ms). An important feature in practical SIES implementation is the capability of its energizing from a source with low electric power. It is obvious that for long application of such storage, additional expenses on cooling will be required; however, the benefit from their application, as a result, will significantly cover these expenses. To participate in daily regulation of the energy flow in the system, the energy of the order of 10^{12} J must be accumulated and the power of several hundreds of megawatt must be produced. Two decades ago the USA begun serial production of SIES with power from 460 to 2500 kVA intended for compensation of voltage rise and decline during 100 ms. Works are continued on the SIES modernization for energy capacity from several units to several hundreds of megajoules to produce energy capacity of 10 MW during 100 s. These storages are quite competitive with storages of other types.

In 1986 materials were synthesized with high-temperature (near 100 K) superconductivity. The synthesis of the materials possessing superconductivity at room temperature would mean break in the spheres of electric power engineering, transport, astronautics, pulsed power, and defense technology.

7.3 Potential and Kinetic Energy Storages

In traditional methods of electric power generation, the primary energy storage (in non-electric form) is placed in front of the electric generator. For example, water stored in a water basin of HPP; coal, gas or black oil of TPP; nuclear fuel of NPP. To compensate for (to level of) the instability of electric energy generation and consumption, storages of energy in non-electric form, as well as storages of electric energy, are placed (inserted) between the generator and the consumer. When the capacity in a system is deficient, they generate additional electric power and hence increase the power.

There are several types of non-electric energy (mainly potential or kinetic) storages used in electric power engineering or promising for such application:

- pumped storage power plants,
- air compression energy storages,
- inertial (flywheel) kinetic energy storages,
- electromechanical energy storages.

7.3.1 Pumped Storage Power Plants

Electric power generation by electric power plants and its consumption by different users are interrelated so that according to physical laws, the electric power consumption at any time must be exactly equal to the electric power generated at that time.

For ideal regular electric power consumption, power plants would be operated under regular loads. Actually, the majority of individual users have irregular electric load schedules, and the total electric power consumption is irregular. Many examples of irregular operation of setups and devices consuming electric power can be given. A plant working in one or two shifts consumes electric power irregularly during 24 h. The electric power consumed by it during non-working time is close to zero. Streets and flats are illuminated only at definite hours. Household appliances, ventilators, vacuum cleaners, electrical stoves, heating devices, TV sets, radio sets, and shavers consume electric power also irregularly. The load of municipal services is maximum one in the morning and evening hours.

The electric load schedule of a region or city is a time history of total power consumed by all customers. It has minima and maxima. This means that at certain hours, a larger aggregate capacity is required, whereas at other hours, a part of generators or electric power stations must be switched off or must operate with a reduced load. The number of electric power stations and their aggregate capacity are determined by a relatively short maximum of electric load of customers. This leads to under-loading of the equipment and increases the cost of power systems.

Thus, a decrease in the operation period of large TPP from 6000 to 4000 h annually causes the cost of generated electric power to increase by 30–35%.

An analysis of mechanisms of electric power consumption demonstrates that its irregularity will further increase with improving well-being of people and an increase in the load from community enterprises in the process of increase in the power supply per production unit and decrease of night shifts. Reduction of working time also increases the irregularity of power consumption. In the majority of countries of Western Europe, the irregularity of power consumption is such that the load can change by 30% of the peaking capacity within one hour, and further increase of the irregularity is expected. It is very difficult to change radically the character of power consumption, because it depends strongly on the human biorhythms and on a number of objective circumstances independent from people.

The power engineers try to balance the schedule of the total load of customers whenever possible. Thus, the cost of electric power depends on the time of day. If the electric power is consumed during periods of maximum loading of an electric power system, its cost increases. This makes customers to be interested in reorganization of their job to reduce the electrical load during periods of maximum power consumption in the electric power system. As a whole, the capabilities of balancing the power consumption are rather low. Therefore, electrical power systems must be sufficiently maneuverable and capable of fast changing the output of electric power stations.

In most countries, considerable portion of electric power (up to 80%) generated by TPP is very sensitive to a regular load schedule. TPP units are badly adapted to power control. Boilers and penstocks of these stations admit load changing only within 10–15%.

Periodic TPP switching on and off does not allow the problem of power control to be solved. In the best case, several hours are required to put a thermal power station into operation. In addition, the operation mode of large TPP with a sharply varying load is undesirable because of increased fuel consumption and higher rates of wear of the equipment and hence reduction of its reliability. We must also take into account that TPP with high vapor parameters have the minimum technically allowable power equal to 50–70% of the rated power of the equipment. The foregoing is true not only of the conventional TPP but also of nuclear power stations. Therefore, the deficit of maneuverable power (peaking load) is covered now and will be covered in the near future by HPP for which the aggregate capacity can be increased from zero to the maximum value in 1–2 min.

The output of an HPP is controlled as follows. When the load in the system is minimum, the HPP has low output, and water falls a storage reservoir. In this case, the energy is accumulated. When the load maximizes, the units of the HPP are put into operation, and their output increases by a desired amount.

The energy accumulation in storage reservoirs on flat rivers leads to flooding of vast territories, which in many cases is very undesirable. Small rivers are unsuitable for output power control in the system, since they have no time to fill the storage reservoir with water.

The above-mentioned problem (removal of peaking loads) can be solved by PSPP operating as follows, Figs. 7.3 and 7.4.

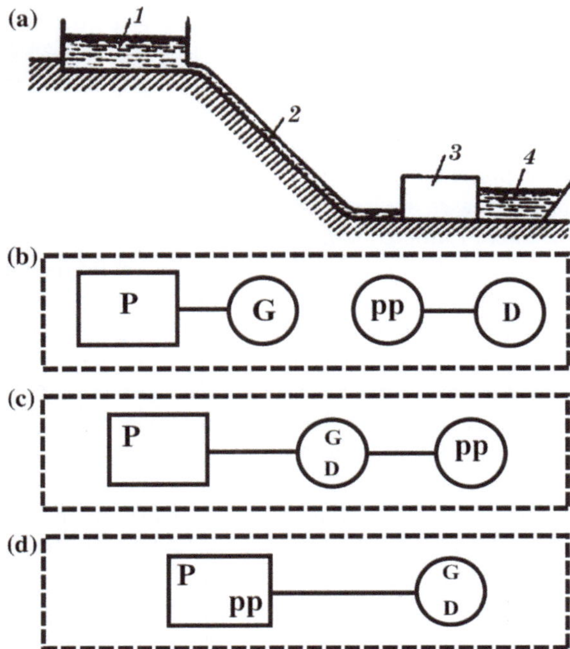

Fig. 7.3 A simplified scheme of the pumped storage power plant: **a** principle of operation; **b–d** three possible options for the layout of the power equipment *P* penstocks, *G* generator, *p* pump, and *D* motor

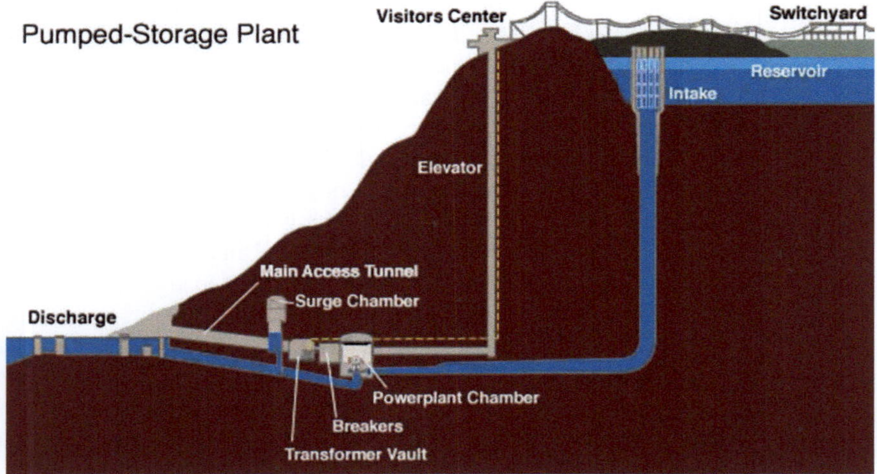

Fig. 7.4 Layout of the main elements of the PSPP

Fig. 7.5 Administrative building and water lines (penstocks) of Zagorskaya PSPP

When the electrical load in interconnected power systems is minimum, PSPP pump water from the lower storage reservoir to the upper one and thus consumes electric power from the system (Fig. 7.3a). In the mode of short peak loading, PSPP generate electric power and spend water stored in the upper reservoir.

Only in the European part of Russia about 200 PSPP can be constructed. In power systems of central, northwest and southern parts of Russia, where the deficit of maneuverable capacity is at maximum, natural slope of the relief allows plants with a small head (80–110 m) to be constructed.

Figure 7.5 shows a photograph of the PSPP near the city of Zagorsk (Moscow region).

Regions with slope of the relief favorable for the construction of PSPP with a head of ~ 1000 m are far from the centers of consumption of maneuverable power. The cost of transmission lines in these cases can significantly exceed the cost of PSPP construction.

The importance of PSPP increases, once high-duty (rated at >1000 MW) nuclear power plants (NPP) have been put in operation, since NPP operate most efficiently with a constant load, and load changes must be balanced by PSPP.

The Lorch-on-Rhein (Germany, 2400 MW), Cornwall (the USA, 2000 MW), and Loch Lomond (England, 1200 MW) PSPP are among the largest ones.

The PSPP capacity depends on the water discharge and head. For a limited volume of the upper reservoir, the capacity can be increased at the expense of a higher head. Therefore, mountainous regions are more convenient for the construction of PSPP. For example, a huge head (about 1800 m) has the Reusseck PSPP in Austria. Natural water reservoirs are used as upper storage reservoirs whenever possible. For example, a high-mountainous pond is used for the Loch Lomond PSPP in England. In Japan, a crater lake is used as the upper storage reservoir of the Numatsavanuma PSPP.

The construction of artificial reservoirs is laborious and expensive. In addition, water can leak from the upper reservoir, and a leak of even several percent reduces significantly the efficiency of HPS. Therefore, careful waterproofing is required.

Water in the upper reservoir can be heated using warm water of condensers of a thermal power station. Closely arranged PSPP and TPP are well combined with each other. PSPP generate electric power at peak hours, the TPP cooling reservoir is used as a lower PSPP reservoir, and warm water from it is pumped to the upper reservoir.

PSPP penstocks through which water from the upper reservoir runs to turbines must not introduce large power losses. The ducted penstocks of large diameters covered from inside with concrete or metal are most convenient. The number of penstocks should be no less than two because in the case of repair of one of them, another penstock can be used. Metal penstocks are less desirable because of large wall friction losses, since these penstocks have relatively small diameters and hence a lot of them must be used.

Penstocks (PS) and generators (G) were used in first PSPP generating electric power, and electric motor (M) and pumps (P) were used to pump water to the upper reservoir (Fig. 7.3b). These stations were called four-machine stations according to the number of machines. By virtue of the independent operation of the generator and pump, the four-machine configuration is sometimes most efficient economically.

For example, generators of the Grimsel PSPP in Switzerland are arranged on one river, whereas pumps are placed on another river flowing above. In this case, the energy spent to pump water to the upper reservoir is less than the energy generated by water running through turbines.

A decrease in the number of machines reduces significantly the SPS cost and opens perspectives for their wide application. Functions of a generator and motor combined in one machine gave rise to three-machine configuration of plants (Fig. 7.3c). A few tens of such plants have been built worldwide. For example, the Hesse Stadt plant rated at 130 MW was built in Germany in 1958, and the Festiniong plant rated at 300 MW was built in England in 1961.

The PSPP became especially efficient after the appearance of reversible hydraulic turbines operating as turbines and pumps (Fig. 7.3b). The number of machines in this case is reduced to two. However, plants with two-machine configuration have lower efficiency because of the necessity to increase the head by a factor of 1.3–1.4 in the pump mode to overcome friction in penstocks. In the generator mode, the head is lower because of friction in penstocks. In order that the unit operated with equal efficiencies in the generator and pump modes, its rotation velocity in the pump mode must be increased. Different rotation velocities in reversible generators complicate their design and make them more expensive.

The perspectives of PSPP application depend in many respects on their efficiency, which for these stations is defined as a ratio of the energy produced by the station in the generation mode to the energy spent in the pump mode.

In the early 20th century, the efficiency of the first PSPP was $\leq 40\%$; the efficiency of modern PSPP is 70–75%. Among the PSPP advantages, low construction cost should also be mentioned. Unlike conventional HPP, it is not required to dam a river, to construct a high dam with long tunnels, etc. HPP on large rivers need 10 m^3 of concrete to be spent to generate 1 kW of installed capacity, whereas large PSPP need only a few tenth of cubic meter of concrete.

PSPP and wind power stations distinguished by unstable operation are well combined with each other. However, it is difficult to rely on the output of wind stations during peak loading of power systems. Given that the electric power generated by wind power stations is accumulated in water pumped to the upper reservoir of PSPP, in the required periods it can be utilized in the electric power system. The lack of reliable methods of electric energy accumulation will promote the PSPP spread.

7.3.2 Air-Compression Energy Storages

This method of energy accumulation was patented in 1949, and the first system was built in 1978 in Germany. In systems of this class, the energy is accumulated in a compressed gas. When the electric power is excessive, the compressor with an electric motor drive pumps a gas in a tank. When it is required to use the stored energy, the compressed gas is supplied into a reversible pneumatic machine with synchronous electric machine operating as a generator or in a turbine or other device directly performing the required mechanical work. There is a considerable number of devices with low capacity directly using the energy of the compressed air (for example, pneumatic weapon, pneumatic brake mechanisms of wheel means of transportation, engines of miner air ducts, etc.).

This technology as applied to the EPS is in the initial stage of development because of the disadvantages inherent in it:

(1) Low efficiency (high losses, up to 30–40%, are caused by gas heating during compression; in the storage regime, heat is transferred to the environment),
(2) Special requirements to the deployment site (presence of underground emptiness of natural or of techno genic origin),
(3) Comparatively large time of power reverse (about 10 min).

Proponents of this technology and power plants based on it (air accumulating power stations—AAPP) emphasize that their use in some cases is reasonable and effective:

- The number of cavities in the earth suitable for creation of AAPP in densely populated areas is larger than the number of rivers, where creation of HPP is possible;
- Some experts show that the construction of AAPP does not damage the environment, but improves it;
- There are projects which allow the application of heat produced during air compression for subsequent utilization. The rise of the air temperature in air

reservoirs of AAPP with gas expander causes its increase to the levels which make AAPP implementation economically efficient not only as an accumulator and electrical energy generator, but also as a source of thermal energy applicable for heating systems;

- There are other AAPP modifications with improved working characteristics. For example, there is the modification that provides application of two-section air reservoir which is to be only once filled by air taken from the atmosphere; the pressure of this air must be different in each of the sections; then the sections are to be isolated from the atmosphere. The compressed air then circulates from the section of lower pressure level to the section of higher pressure level—with the help of a compressor, and from the section of higher pressure level to the section of lower pressure—passing through the gas-expander connected to the shaft of the electric power generator, Fig. 7.6.

The proposed methods of increasing the efficiency of air-compressor energy storage at the current level of technology are feasible mainly in low-power systems. The author of this textbook found in the publications a description of only two

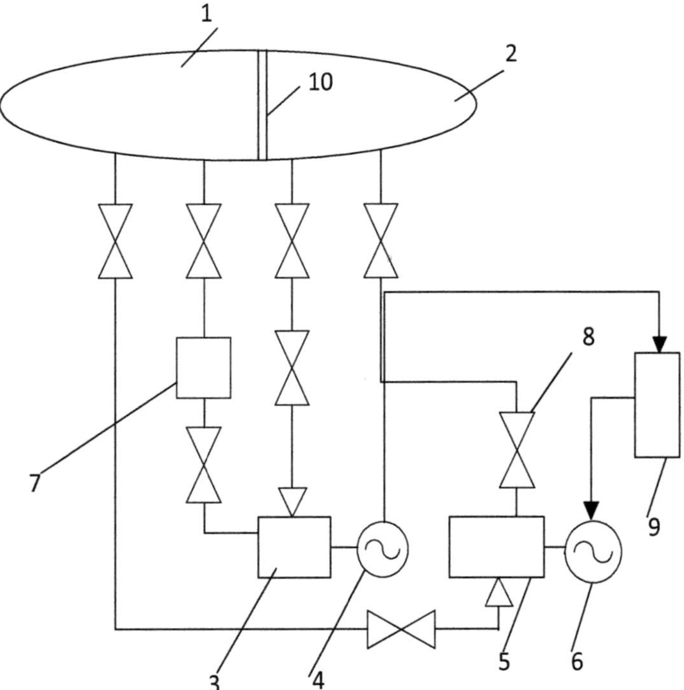

Fig. 7.6 Functional diagram of AAPP: *1* high-pressure air reservoir (ARH), *2* low-pressure air reservoir, *3* gas expander, *4* electric generator, *5* compressor, *6* compressor electric drive, *7* pressure regulator, *8* locking valve, *9* electric power system, and *10* partition

large-scale pilot systems based on injection of compressed air into underground reservoirs.

7.3.3 Inertial Energy Storages (Flywheels and Super-Flywheels)

Inertial (dynamic) energy storages (IES) operate as follows: the engine drives a flywheel—the flywheel accumulates energy—the generator rotated by a flywheel generates electric energy. Their practical application begun in the end of the XIXth century for short-term energy production in various technical systems (sea torpedoes, rolling mills, and melting furnaces). From the middle of the XXth century, their application has significantly increased in connection with the development of new types of weapon (beam, electro kinetic, and other similar weapons) with the development of high-energy physics. IES with cyclopean overall dimensions have been manufactured having the characteristic parameters: flywheel (steel) diameter of 3–5 m, length up to 6 m, mass of ~ 200–250 t, output (discharge) power of several tens of megawatts, power capacitance up to 10 MW·, and flywheel rotation velocity 1000–1500 rpm have been manufactured.

Works on IES modernization have led to several original technical solutions on reduction of energy losses: the flywheel is placed in the environment of light gas (hydrogen or helium) or in vacuum; gas static, gas dynamic, or magnetic supports (bearings and toes); flywheels of complicated design are used, and composite fibrous materials are used as a flywheel material rather than steel, Fig. 7.7.

Such flywheels were called super-flywheels, since they possess a number of advantages over steel ones, first, high specific power capacity (5–15 MJ/kg or 1.4–4.17 kWh/kg) practically unattainable for all known storages. This is provided, basically, by two factors: (a) high mechanical strength of the flywheel allowing to drive it to 7000 rpm and even greater and to accumulate a considerable amount of the kinetic energy, and (b) capability of accumulating potential energy in the form of elastic flywheel deformation. Modern superfly wheels are manufactured from nonmetallic fibers with low density and high mechanical strength (organic fiber—kevlar, fiberglass, and carbon fiber). Fibrous materials are elongated under the stretching action of centrifugal forces in the rotating flywheel and reserve, except the kinetic energy, the potential energy. The amount of the latter is determined by relative deformation of fibers and deformation can be commensurable with the kinetic energy for high ratio of the admissible pressure to the elasticity modulus (to the Young elasticity modulus) for the flywheel material. Such flywheel can deliver considerable portion of the total stored energy for comparatively small change of the angular velocity during deceleration, i.e., in the discharge regime.

The IES intended for application in the EPS to level peaks of power consumption, must work together with the reversible electric machine mounted on the shaft common with the flywheel. The electric machine in the charge regime is

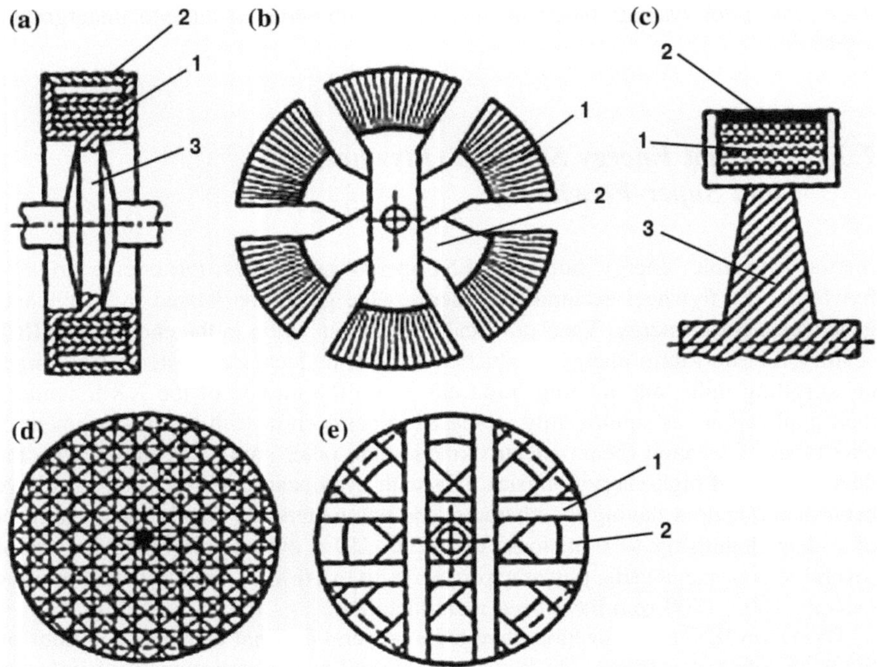

Fig. 7.7 Designs of super-flywheels: **a** ribbon (comprising metal ribbon *1*, casing *2*, and light disk *3*), **b** rod (comprising fiber cores *1* and holder *2*); **c** rim (comprising fiber rim *1*, bandage *2*, and light disk *3*), **d** disk (quasi isotropic composite design from fibers), **e** rim-disk (comprising fiber rim *1* and ribbon spokes *2*)

energized from an alternating current system (50 or 60 Hz) through a frequency multiplier and spinning the flywheel until the rated velocity. In the discharge regime, this machine is rotated by the flywheel and operates as an electric energy generator delivered to the EPS. The time of power reverse—transition from the charge to the discharge regime—takes a few tenth of a second. The IES projects (as a rule, vertical configuration) for energy from 70 to 200 GJ with a set of disk flywheels having diameters from 4.5 to 18 m, total height from 6 to 10 m, with rotation velocities up to 7200 rpm have been developed.

It is considered that the application of the IES in the EPS will be expedient if they have the following characteristics:

(1) Power capacity greater than 10^3 MWh,
(2) Specific capital investments less than 1000 dollars/kW,
(3) Service lifetime about 30 years,
(4) Specific power capacity no less than 3.5 kW h/m^3,
(5) Efficiency no less than 80–85%.

At present, they are too expensive because of high cost of materials and complexity of construction of such storage system. For example, because the angular flywheel velocity is much greater than the velocity of electro-generators, transmission devices are required that not only complicate and make more expensive the design, but also cause additional energy losses. The same effects will be caused by application of several tens of flywheels of smaller sizes and more technological rather than one huge flywheel (its mass would reach 1000 t). Special protective measures from possible destruction (rupture) of the flywheel should also be provided.

7.3.4 Electromechanical Drive Storages

As a rule, in the electromechanical storage (EMS) the kinetic energy of the flywheel and rotor of the electric machine (EM) connected to it is stored. In some EMS there is no flywheel as the element independent or structurally combined with the EM rotor, and functions of energy storage are fulfilled by the rotor (along with the functions in the electromechanical EM processes). As a generator, the same EM, which serves as the electric motor for flywheel drive, is used or another EM also mounted on the EMS shaft. The EMS represents not only the storage, but also the power amplifier (energy compactor). In the charging regime, the EM engine consumes a rather small power from the power supply, and in the mode discharge, the EM generator delivers much higher power for a shorter time. Depending on designation, the EMS can be constructed based on EM of various types: alternating current (synchronous or asynchronous) and direct current ones, including collector or opposite and unipolar valves.

The block diagram of the power complex with EMS is shown in Fig. 7.8.

The EMS of two types: (1) synchronous machines with frequency converters in the primary circuit and with flywheels on the shaft and (2) asynchronous machines with flywheels on the shaft have found the widest application in electric power engineering.

Nowadays there are no practical restrictions on the creation of units of the first type with capacity to 300–400 MW and of the second type with capacity to 800–1600 MW. In the pulsed (forced) regime the commercially available synchronous electric generators used in electric power stations, can produce a capacity up to 10^9 W for several fractions of a second. In this case, up to 10 MJ per pulse is supplied to a load with efficiency of kinetic energy supply from the flywheel of 2–3%. The synchronous impact generator can produce 1000 MJ pulses of power up to 100 MW. By the stored energy level, the synchronous generators can generate a pulse train (with period between pulses of several minutes required to restore the frequency of rotation of the rotor). The exploitation of synchronous generators is constrained mainly by a low efficiency of energy transmission to a load.

In the last decades the development and implementation of the Smart Grid concept and the increase of the fraction of the distributed electric energy (including

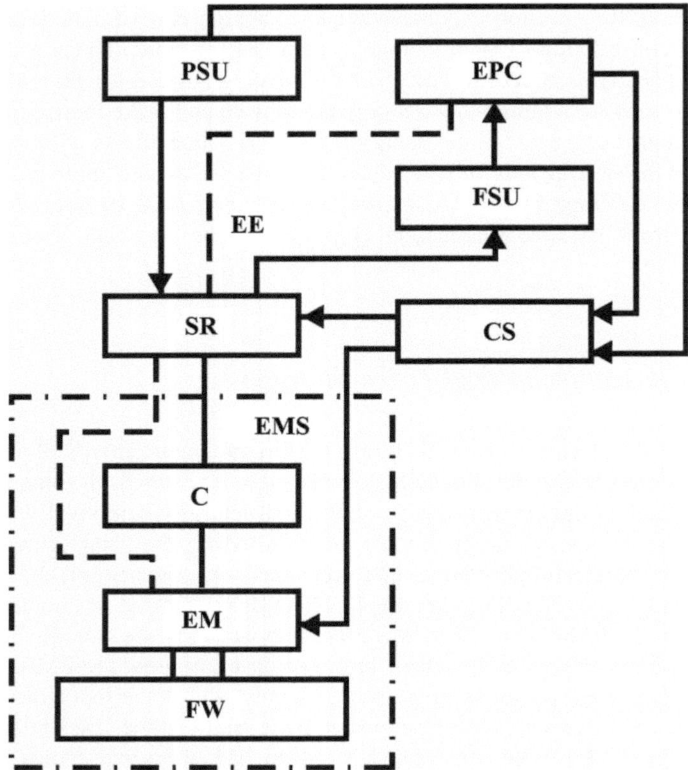

Fig. 7.8 Block diagram of the power complex with the EMS comprising power supply unit PSU, charge-discharge regime switch SR, reversible commutator C in the EMS with direct current EM, electromechanical brush-collector unit with the EM of classical design, and controllable semiconductor converter with the contactless valve EM; here *EM* is the electric machine; *FW* is the flywheel; *EPC* is the electric power consumer; *FSU* is the frequency stabilization unit (applied if the EM is used as an engine or alternating current generator and the block C is absent; see the *dashed connection line*); and *CS* is the control system

on the basis of URPS) in the electric power balance as well as the increased severity of consequences (economic, social, etc.) of interruptions or undersupply of electric energy have increased the importance and role of the energy storages.

Questions and Tasks to Chapter 7

1. Name the basic ways of electric and non-electric energy accumulation.
2. What are the main functions of the energy store (of the accumulator) in electric power systems? What are the positive effects in the electricity that are provided by the use of energy storage?
3. What are the most important parameters of energy storage devices when they are used in the electrical power engineering?

4. Fuel elements (cells): principle of operation, designation, advantages, and disadvantages.
5. Water storage power station: principle of operation, advantages, disadvantages, and the current state.
6. Types and design of inertial energy storages. How can you describe the work of the inertial energy storage?
7. Ways to improve operating parameters of inductive energy storage.
8. What interferes with large-scale application of air-compressor energy storages in electric power engineering?
9. What are strong points and weaknesses of capacitive energy storages?

Reference

1. Meadows D, Randers J, Meadows D. The limits to growth: the 30-year update. White River Junction, Vermont; 2004.

Chapter 8
Power Engineering and the Biosphere

8.1 Power Engineering as the Threat to the Biosphere

As noted above, the world in its evolution has reached the stage that calls for the development of new global approaches to a solution of the most complicated current problems. Among these problems is "ecological infarct." Under these conditions, of primary importance is prediction of the mankind future, associated with the state of the environment, to understand contradictions it will face and measures that will have to be undertaken to make them less acute as well as limits in which this will be achieved and at which cost. For many decades, the annual increment of energy production has been about 3%. If such rate of energy increase is retained, the allowable heat limit will be attaining in 60 years. Then by the middle of the 21st century, the production of these types of energy will have been stabilized. In this case, a difficult question arises on quotas of energy production for each state. This will initiate a global crisis of our civilization that will be of geopolitical importance.

Actually, the power of only stationary power plants all over the world is about 1.9 TW, and the power of all systems producing energy is no less than 10 TW, which is approximately equal to powers of such natural phenomena as moisture vaporization from the Earth's surface (0.5 TW), sea and ocean tides (2–3 TW), thermal gradients in oceans and dry land (2–2.5 TW), and even of such colossal forces of nature as earthquakes (1.5–100 TW). The energy of anthropogenic origin influences natural processes proceeding on our planet. The energy of all fuels extracted from the planet per year is ~ 60 thousand TWh. It makes less than 0.015% of the solar energy reaching on the Earth's surface, but even for such value of energy produced, its thermal effect on the heat balance of our planet is noticeable. This energy influences the climate, especially in most "energetically stressed" territories.

© Springer International Publishing AG 2018
V.Y. Ushakov, *Electrical Power Engineering*, Green Energy and Technology,
https://doi.org/10.1007/978-3-319-62301-6_8

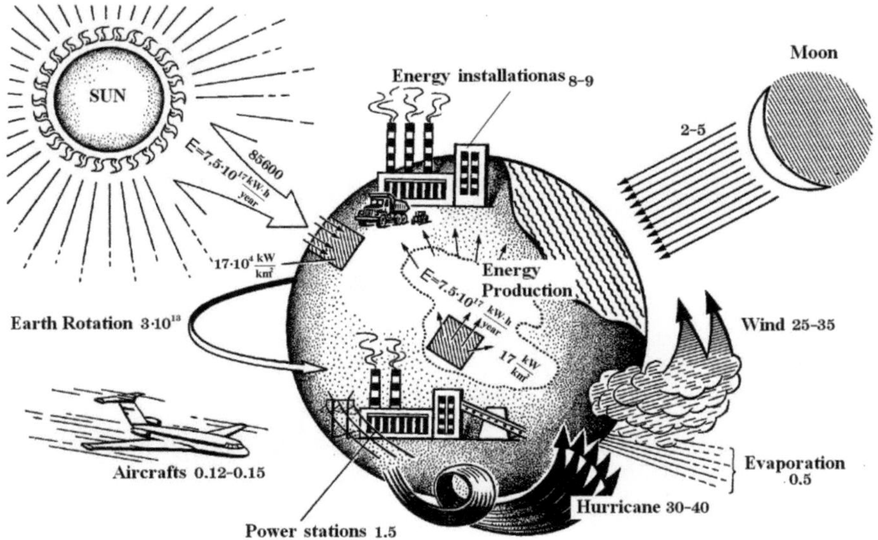

Fig. 8.1 Powers and energy of some geophysical processes and anthropogenic power systems

Table 8.1 Comparison of the power of energy floods natural and anthropogenic origin

Natural and anthropogenic phenomena and processes	Power (TW)	
	Current	In the middle of the XXIst century
Anthropogenic energy	12.0–14.0	55.0–100.0
Power of all power stations	4.8	25.0–40.0
Thermal potential of oceans and land	2.0–2.5	
Floods and tides	5.0–6.0	
Hurricanes (tornado)	20.0–30.0	
Earthquakes	25.0–40.0 and more	

Figure 8.1 and Table 8.1 show the relationship between powers of anthropogenic systems of energy conversion and powers of natural origin expressed in watts.

Practical and economic human activities were historically developed so that, as a rule, their ecological consequences on the environment were disregarded. This was allowable until their scales and power resources brought into action were insignificant compared to powers of environmental phenomena.

In some cases, unfavorable changes were observed in local regions. Thus, for example, when forests had been cut and soil had become infertile because of irrational agricultural management, the human activity could be organized in another place.

Nowadays people feel the consequences of sometimes extremely unfavorable effects on the environment already in their everyday life. Intense pollution of the atmosphere by harmful substances changes its gas composition, forest areas fast decrease, the ecological balance between the atmosphere and the global ocean is disturbed due to contamination of ocean water by oil and other products of human activity, and the so-called thermal pollution of water and air heated by various industrial wastes is observed [1].

Anthropogenic emissions are particularly dangerous in that they **reduce the stocks of drinking water** and cause a **dangerous rise in temperature of surface and the atmosphere** because **of the greenhouse effect** and **thermal pollution.** The pollution of the atmosphere, hydrosphere and lithosphere causes a decrease in the quality of drinking water of open sources (rivers, lakes, ponds, artificial reservoirs) and underground water too. Groundwater sources are, in the opinion of hydrogeologists and environmentalists, the last hope of people to meet their needs for drinking water in the near future. Per capita of water consumption, like the energy consumption, characterized by a large difference between water consumption in different countries and regions. Already more than 1 billion people live in conditions of shortage of drinking water.

Since humanity has entered into the seventh environmental crisis (now it has the global character), we are gradually drawn into technical (technological) revolution. For power engineering it means the development of new energy sources and new ways of converting them into electrical and thermal energy. The beginning and pace/rate of this revolution depend upon a number of factors:

1. Economic—requires investment on a huge scale
2. Technology—inability of quick change of technological structures
3. Social:

   (a) rapid population growth in developing countries, and
   (b) inability to quickly change the existing mode of life

4. Global (historical and geographical)—extremely uneven distribution of energy resources across countries and regions.

At present, these problems have grown up to scale that rises a question whether the scientific and technical progress threatens to destroy the civilization because of its harmful effect.

A power system in which electrical and thermal energies are produced is directly connected with a system providing primary power resources (Fig. 8.2). There are also feedbacks of various factors of human activity with the state and development of power engineering. Exactly for this reason, power systems should be considered as subsystems of an integral global system of functioning of the human society.

Creation and preservation of a power system and conditions of its operation in many respects are determined by natural factors, for example, by the availability of water reservoirs and geographical location of power resources and customers. A state of the biosphere and a degree of its pollution connected with the operation

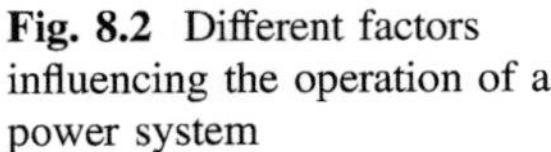
Fig. 8.2 Different factors influencing the operation of a power system

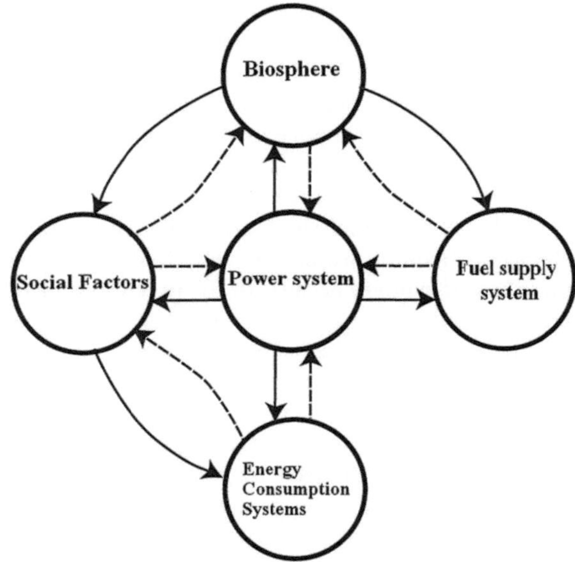

of power systems impose definite limitations on their characteristics and working conditions. Direct relationships and feedbacks between the biosphere and a power system are traced here.

Efficient management of a power system must consider its influence on the biosphere, social functions of a system of fuel supply, demands for energy of industry and transport, and other factors. Not only energy production in the power system, but also its consumption by various branches of economy must be managed.

Modern power systems, as indicated above, are closely related with many-sided human activity. They influence directly a great diversity of branches of economy (industry, transport, and agriculture), economics, social conditions, state of the biosphere, etc.

Interactions of power systems with the related economic system and the biosphere must be considered in the process of managing the power system, predicting and designing its future development, and carrying out scientific research. The development of power engineering influences directly the general state of technology. In turn, achievements in different branches of technology are reflected in the state of power engineering. The power engineering, the biosphere, and the social and economic human activities are interrelated.

The most important difference between the two above-discussed types of power sources (renewable and non-renewable) is their effects on the biosphere. Power engineering based on non-renewable sources results in additional heating of the habitat, that is, the energy of these sources is added to the energy of heating of our planet by the Sun.

Renewable power sources do not heat our planet; therefore, this type of energy is called "non-adding" or "green." Indeed, in this case, using, for example, solar rays in power systems located on the Earth, we eliminate their energy from the heating cycle of the planet. After usage, we return this energy to the planet as the same amount of heat. The same is true for the wind or oceanic energy, namely, as much energy is returned to the habitat in the form of heat as is taken from the power background. The non-adding energy can be called *waste-free*, whereas the adding energy should be considered as the energy *polluting the habitat*.

Calculations demonstrate that the production of only 1% of the adding energy (for example, chemical, nuclear, or thermonuclear energy) to the total solar energy will cause the average temperature of the biosphere to increase by $\sim 1\,°C$.

It is generally recognized that such an increase in the average temperature of the biosphere will cause disastrous global consequences not only on the Earth's geography and climate, but also on animal and vegetable kingdoms.

Discovery of processes in atmosphere, which was made by means of satellite, has shown that water vapor produced in increasing amount due to the increase of temperature destroys ozone layers. As is well known, ozone layers protect people and animals from ultraviolet radiation.

Increased threatening changes in the nature are illustrated in Figs. 8.3, 8.4, 8.5, 8.6, 8.7, and 8.8. Over the last century the atmospheric temperature has risen by 0.74 °C (Figs. 8.3, 8.4, and 8.5), and the level of the oceans annually increased by 3 mm (Fig. 8.6).

Global warming will unpredictably change the rate and character of all processes proceeding on the Earth and will considerably change the atmospheric circulation and soil moistening. As a result, zones optimal for agriculture and other types of economic activities will be strongly displaced.

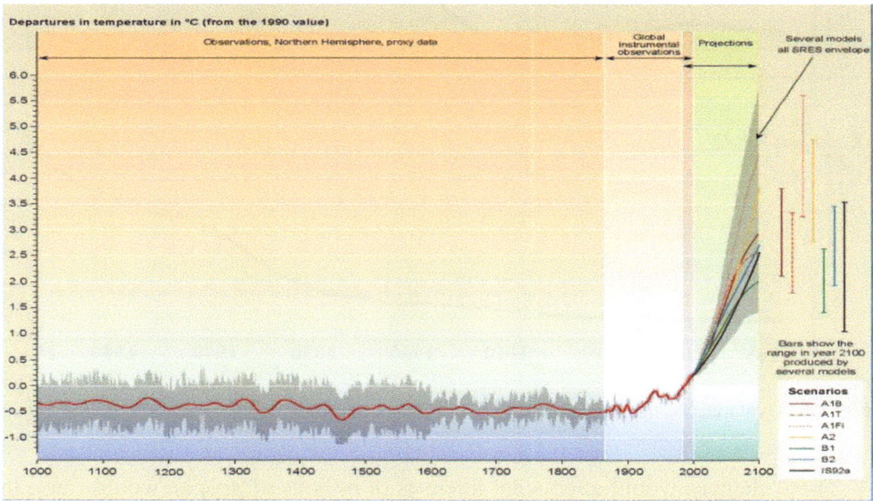

Fig. 8.3 Variations of the Earth's temperature: from 1000 to 2100 (evaluations and measuring)

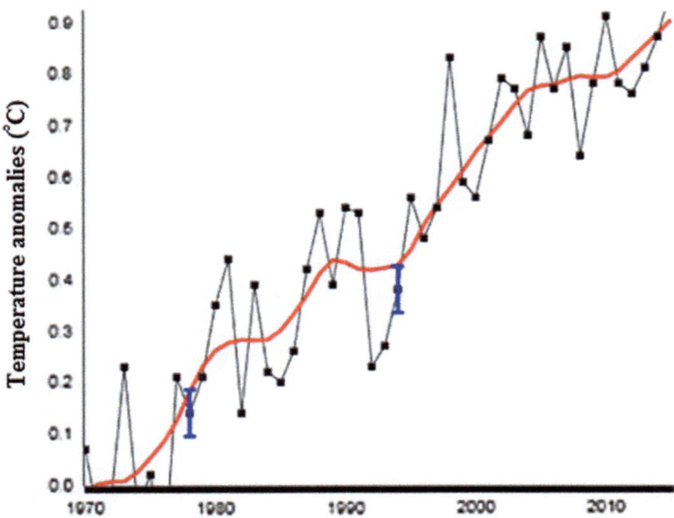

Fig. 8.4 Global temperature changes over the past 40 years (ground measuring)

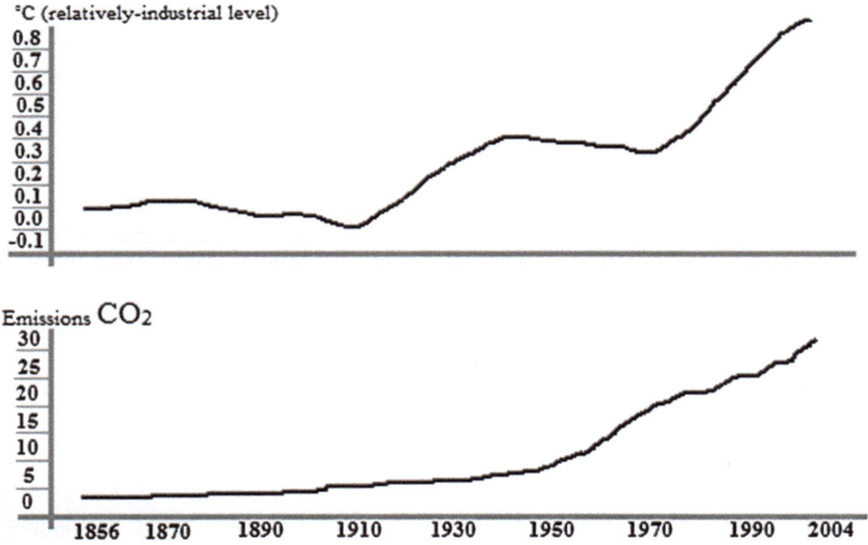

Fig. 8.5 Dynamics of the temperature rise and CO_2 emissions

These changes would affect vital interests of billions of people, would cause the massive migration of population (including crossing of national boundaries), the resettlement of people to zones without developed infrastructure and, as a result,

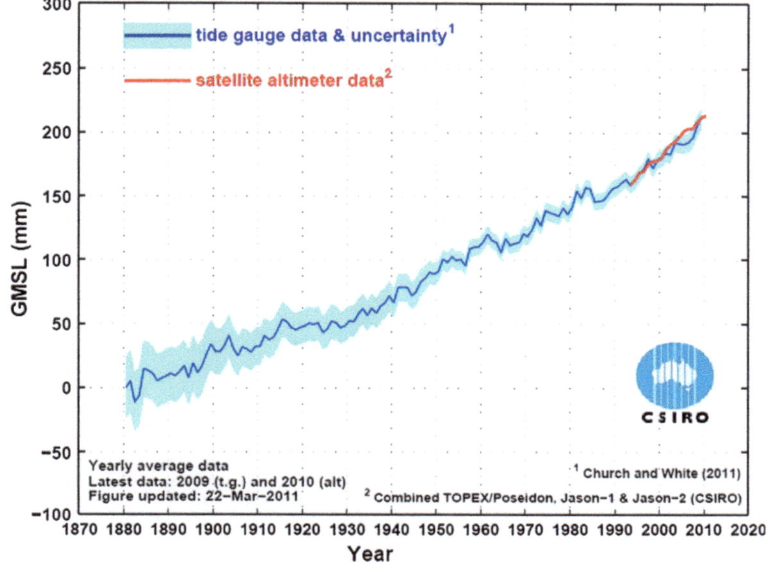

Fig. 8.6 Global mean see level during 1880–2010

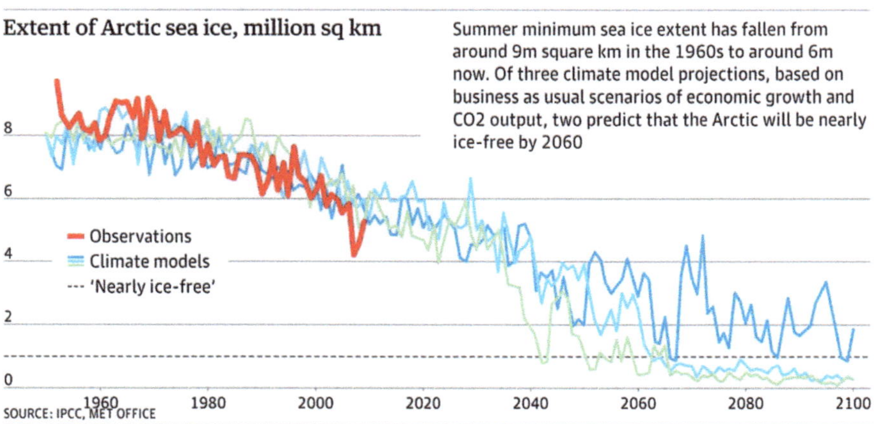

Fig. 8.7 Observation and climate models relatively of Arctic sea ice extend

disastrous social consequences. It is especially unpleasant that all these changes will occur very fast on the historical time scale, and it will be very difficult to adapt to them.

Over the past few years, the problem of global *greenhouse effect* has attracted a close attention of the scientists.

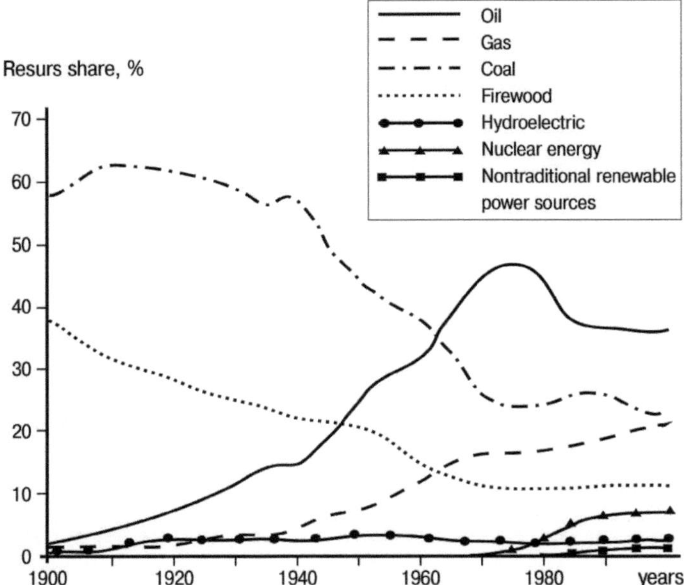

Fig. 8.8 Evolution of the world energy consumption structure

Unfortunately, it turns out that it is very difficult to predict the future correctly because of a high degree of uncertainty of this problem due to our poor knowledge of many processes proceeding in the atmosphere, soil, and ocean. It is not surprising that predictions of different scientific groups differ radically from each other. An optimistic prediction implies that the temperature will grow by 1 °C as the amount of carbon dioxide is doubled, whereas a pessimistic prediction implies that the temperature will grow by 5 °C. An average value of 2.5–3 °C is taken as the most probable value of temperature rise. Predicting the performance of these processes is complicated by the fact that the temperature on the Earth depends on the little-known processes of cosmic scales.

How fast will carbon dioxide accumulate actually? If the rate of carbon dioxide increase retained, in 50 years the carbon dioxide concentration will increase by 15–20% compared to its value at present. However, since the consumption of fossil fuel grows, it seems more probable that 25% and more by the early 40s will increase the carbon dioxide concentration. In the context of the most probable prediction, this would correspond to a temperature rise of $\sim$1 °C, which would be very dangerous.

Most likely, a safe limit of adding energy will not exceed one tenth of percent of the power of solar energy incident on the Earth's surface, that is, will be about 100 billion kilowatt. Nowadays the Earth's civilization produces an adding energy power of 10 billion kilowatt for consumptive use (in industry, everyday life, and transport), that is, only 10 times less than the allowable limit.

The population growth and the intensity of its activities to improve the quality of life reinforce the technological impact on the environment thanks to changes in the following areas:

(1) structure of the earth's surface (plowing up of steppes, deforestation, land reclamation, creation of artificial lakes and seas, and so on);
(2) composition of the biosphere, circulation and balance of its constituent substances (removal of mineral resources, creation of dumps, and emissions of various substances into the atmosphere, ground and water);
(3) heat balance of certain regions of the globe and the planet (heat emissions from fuel combustion—thermal pollution, "green house" gases (GHG), and so on);
(4) biota (the extermination of certain breeds of animals and varieties of plants, moving them to new habitats).

In contrast to natural fluctuations, the anthropogenic impact leads to sharp, rapid changes in the state of the environment in the region. As part of the natural environment, new components are characterized by the term "pollution." Nature responds to these effects increasing the instability and unpredictability of weather, including in the form of negative phenomena. Over the past 30 years, the annual number of natural disasters has increased by 4 times, and the economic damage from them—by 7 times.

There are scientists—climatologists (they are in the minority), which explains the direction and causes of climate changes in a different way. Proponents of the alternative point of view call "Global Warming" "the largest political-scientific fraud in the history of mankind."

Their arguments are:

- No warming on the Earth has been observed already for about 20 years; the warmest year in the last decades was 1998; since then the average temperature of the planet has not increased.
- Abnormally cold winters from which the Europe and even North Africa have suffered affected in the last few years are perceived dy a number of experts as symptoms of future global cooling.
- Warming in the 90s was connected which the coincidence of peaks of two— 200-years and 60-years natural of "warming—cooling" cycles.
- Cooling in the 40–70-ies of XX century was observed during the period of rapid growth of industry and emissions of greenhouse gases.
- Near-ground measurements of temperature are incorrect, because many meteorological stations are located in big cities or near them; thereby, they overestimate the indications of thermometers.
- Facts of deliberate information distortion or its hiding to reinforce the hypothesis of global warming have been established.

According to the opinion of climatologists, the share of anthropogenic influence on climate is 25%, natural (space) processes making up the rest/In our climatic system, 60-year cycles are observed. The phase of warming has started in the

1970s; nowadays it is coming to an end. The temperature rise similar to that observed at present was already observed 1700 and 1000 years ago when the mankind could not influence the climate. Future climate changes will be influenced by these two factors. At present, an increased number of accidents, which all us observe, accompanies the temperature rise.

8.2   Short Analysis of the Trends in Electric Energy Generation and Consumption in Aspect of the Influence on Environment

First, we must bear in mind that the share of energy accounts for about 50% of all emissions into the atmosphere, hydrosphere, and lithosphere.

Consumption of power resources grows rapidly due to a continuous increase in the world production. According to the data of the International Institute for Applied Systems Analysis (IIASA), the consumption of primary energy will have been increased to ~ 24 billion ton of standard fuel annually by 2030, that is, it will double compared to 1988. The annual increment of primary energy consumption is $\sim 1.5–2\%$. The remaining reserves of oil and natural gas are sufficient for the coming 50–100 years. Of course, these data are approximate; nevertheless, they give us an insight into the future. Table 8.2 summarizes the data on the global consumption of the most important energy carriers in 1990 and the corresponding data predicted for 2020.

Table 8.2 Optimistic and pessimistic variants of the development of global power engineering published by IIASA in 1993

Item	Data for 1990	Prediction for 2020	
		Optimistic variant	Pessimistic variant
1	**2**	**3**	**4**
Population, million people	5292	7518	8092
Economic growth			
Gross domestic product (trln US dollars)	21.0	64.7	55.7
Gross domestic product per capita (US dollars)	3972	8001	6884
Needs for primary energy resources			
Net (million ton of standard fuel)	12,593	24,610	16,120
Specific (ton of standard fuel per capita)	2374	3060	1988
Needs for electric power (billion kWh)	11,608	23,000[a]	23,000[a]
Energy content of economy, kg of standard fuel/dollar	0.55	0.41[a]	0.41[a]

(continued)

Table 8.2 (continued)

Item	Data for 1990	Prediction for 2020	
		Optimistic variant	Pessimistic variant
Structure of the global energy balance (in % to total)			
Coal	26.3	28.2	18.9
Oil	31.0	26.7	25.7
Natural gas	19.5	21.2	22.1
Atomic energy	5.0	5.7	6.1
Water power	5.3	5.8	5.9
Renewable power sources	12.9	12.4	21.3
Regional needs for primary energy resources (million ton of standard fuel)			
North America	3095	3494	2615
Latin America	825	3190	1869
Western Europe	2091	2594	1886
Central and Eastern Europe	418	515	379
Commonwealth of Independent States[a]	2069	2394	1830
Middle East and North Africa	453	1853	1131
Africa to the south of the Sahara desert	380	1829	869
Pacific[b]	2635 (1358)	6989 (3328)	4273 (2528)
South Asia	637	2648	1287
Emissions into the atmosphere			
Sulfur (million ton)	64.6	98.1	42.8
Nitrogen (million ton)	24.0	37.9	20.9
Carbon (million ton)	5.9	11.5	6.3

[a]Intermediate variant
[b]Including Asian countries with planned economy (the data for this group of countries are given in the parentheses.)

Obviously, planning and designing power systems as well as their subsequent development and maintenance must consider all their *effects on the environment*. In this connection, experts in power engineering must have extensive knowledge of the nature and natural phenomena.

Under conditions of a rapid growth of electric power engineering and many other engineering branches when problems of pollution of the biosphere are of great importance, renewable power sources attract much attention (see Chap. 5). Of interest is to trace the consumption of different energy resource types since the beginning of the XXth century, Fig. 8.8.

The muscular energy of the human and animals, sometimes called biological energy, was the sole energy source in former times. Today its share is less than 1% of the total energy consumption (it is not shown in Fig. 8.8). The share of muscular energy will further decrease. This demonstrates that the high level in the development of productive forces has allowed the human to throw almost completely efforts on manufacturing essential products using machines. In order that machines do this work, the human, based on laws of nature that he perceived and implemented in practice, should set to work huge powers applying them to instruments of production. These powers of modern instruments invariably exceed the maximum power of biological sources.

The first heat sources were various organic remains and wood. For a long time, wood was the main energy carrier. Later on, as other energy sources with higher energy content (coal and oil) were mastered, the consumption of wood decreased. Utilization of wood can be stopped almost completely in the near future. Early in the 20th century, the greatest share (of the order of 70%) of all utilized energy resources fell on coal. As the consumption of oil, gas, and electric power increased, the share of coal decreased, though the total amount of extracted coal significantly increased.

A model of the power engineering development was constructed in 1992 (by I. A. Bashmakov), which allows one to forecast the needs of six different energy sources, six secondary energy carriers as applied to six sectors of economy. The starting point of an analysis was a scenario based on the common sense. Then the capability of accelerated addition of oil, coal, gas, nuclear energy, and renewable energy sources to the global energy balance was estimated together with the feasibility of resolving contradictions arising in global power engineering. One more modification was added to six modifications of the basic scenario, namely, a scenario with accelerated energy saving. Table 8.3 gives the global energy balance predicted for 2020 based on this energy balance.

The evolution of the debit side of the global energy balance is determined by two opposite tendencies: (a) gradual approach of the structure of power resource production to the structure of available reserves and (b) improvement in the quality of consumed energy resources and energy carriers.

The first tendency determines an increasing share of coal in the global energy balance. The second tendency determines increasing shares of oil, gas, and nuclear fuel in the global energy balance.

The increased consumption of primary power resources will decrease shares of three main organic fuel types, but they will dominate in the structure of the global energy balance until the middle 21st century and even later.

Undoubtedly, a rational combination of different energy resources and a smooth development of power engineering would allow difficulties, sometimes disastrous in character analogous to those observed in 1973–1974 in a number of countries, to be avoided. These difficulties called the *power crisis* were due to many-year predatory utilization of national raw resources by international monopolies.

Despite of all the difficulties to meet the growing electricity needs of each individual and the increasing world population, in the first place among the problems facing humanity, is environmental protection.

Ecologists have come to a conclusion that "... the mankind has approached a certain critical limit, having faced in their development new external borders ... these borders are defined not only by subsurface area resources or accessible energy sources, but the number of biosphere possibilities aimed at neutralization of growing anthropogenic pressure."

Table 8.3 Global energy balance

Sector of economy	Power resources								
	Coal	Other solid fuels	Oil	Gas	Renewable sources	Nuclear	Electric power	Heat	Total
Primary energy consumption	4951	1344	4054	3823	1461	1376			18,763
Electric power generation	3079	117	406	1102	1461	1376	2728	713	4099
Power sector	140	17	328	403	0	0	429	73	1390
Supplied energy consumption	1732	1478	4788	2318	0	0	2300	657	13,274
Industry	1175	327	557	1103	0	0	1223	401	4787
Transport	50	13	2888	41	0	0	85	4	3082
Public utility sector	501	1128	618	979	0	0	992	251	4469
Non-energy needs	7	9	724	195	0	0	0	0	935

The basic scenario up to 2020 (million ton of standard fuel)

As discussed in Chap. 5, humanity hopes to preserve the ecological well-being in the world are connected with the transition to a "clean" energy—primarily based on the use of renewable energy sources.

No harmful emissions and excessive heat during operation of such sources of electricity and heat help one fully compensate for anthropogenic impact on the environment during their manufacture. For example, projected for 2050 electricity production by means of SPP will help one to prevent annual emissions of 6 billion tons of greenhouse gases (CO_2 and about ten other harmful gases). In the production of solar cells, the level of contamination does not exceed the permissible level for the microelectronic industry enterprises in the production of components for thermodynamic SPP, hot air, or thermal circuit for general engineering company.

8.3 Main Sources of Threats to the Environment in Different Sectors of Fuel and Energy Complex

According to estimations of international experts, up to 40% of the total gross domestic product in developed countries connected with extraction, processing and consumption of energy resources. In the last decades, even non-professionals have realized that industrial-type economy dominating in the world does not adequately consider ecological and social consequences of economic activities; it also produces a lot of problems and dangers threatening human civilization. Ecological threat began to emerge faster than the positive changes in the energy balance (consecutive transition to fuel with a lesser content of carbon or not containing it at all) and in the environmental protection activity.

The experience has proved that in terms of fuel and energy complex (FEC) improvement of ecological basic technologies indicators for ecological safety concedes in struggle to other approach—increase in the efficiency of power resource use at all stages of their service life cycle—from extraction to consumption of electric and thermal energy. It is a more radical solution, since it is aimed at elimination of reasons (power consumption growth), instead of consequences (increase in ecological damage because of energy generation growth). Energy saving character is relevant to economic reorganization and objectively caused by (economic and ecological factors) quality of the present stage of world economy development. Correlation between amount of investments into energy saving actions and expenses on corresponding quantity of energy generation are estimated as 1: 5–8. Goals which are put forward by energy saving are getting of paramount importance among obligatory actions of environmental protection activity.

The World Commission on Environment and Development has developed the concept of the sustainable development, which has received the official status within the frame of the UN as "strategy of world development" in which necessity of searching new interaction society model, economy, and the state interaction model is proclaimed. In the report of Intergovernmental Committee on United

Nations Climate Change "Struggle against climate changes: human solidarity in a divided world" the unequivocal and well given conclusion contains: climate changes occur because of the human being activity. All over the world energy sector emissions of CO_2 in the atmosphere will increase from 23.9 billion tons in 2001 to 37.1 billion tons in 2025 if, of course, extraordinary measures are not undertaken for environmental protection.

There are three basic ways of decrease (or at least reduction of rates of increase) in pressure on the environment of growing energy consumption.

1. Change of balance towards increase in the share of more non-polluting primary energy sources: fuel with smaller carbon content or not containing it at all: uranium, renewable energy sources, and in the long term—fuel for implementation of controlled thermonuclear fusion. It is a perspective way of prevention of ecological accidents, but as the experience of the advanced countries has proved, replacement of traditional hydrocarbon fuel kinds by alternative and non-polluting energy sources demands huge capital investments.
2. Increase in ecological cleanliness of extraction technologies, primary processing and transportation of conventional power sources and generation of electric and thermal energy.
3. Reduction of CO_2 emissions in the atmosphere by burying them in the underground storages. (About 12 power stations have been equipped with such storages in EU by 2015.)

8.3.1 Activities in the Resource Sectors

The greatest damage to the nature of the activity of enterprises of the FEC resource sectors inflicts extraction and transportation of oil and natural gas. This damage is expressed as follows:

1. Pollution of all spheres of the Earth that affects the development of living matter: the atmosphere, hydrosphere, lithosphere, and the earth's surface.
2. Impact on the deep layers of the earth (10–15 km): pollution of underground drinking water and reduced in situ pressure, which can cause earthquakes and soil subsidence.
3. The output from agricultural land about few percent of the total area of farmland is lost annually (a huge number of wells—hundreds of thousands, extended permanent and temporary roads, rail and waterways, pipelines, and overhead transmission lines).
4. A huge number of vehicles producing polluting fuel, oil, and exhaust fumes.
5. Large volumes of water consumption for technological, transport, economic—household and firefighting—needs, and the same volume of discharge of polluted wastewater.

Considering condition of industrial equipment installed in many oil companies, it is possible to ascertain that throughout the next years the problem of *oil* emergency floods liquidation will remain urgent. Prevention of emergencies is considered the most economically sound solution: prevention oil spills due to pipeline damages, automated control of pipeline condition and pipeline sites where pipelines need replacement or capital repairs. Such activity seems to be economically proved in most cases. Land restoration is considered one of the important activities, which should take place after field facility liquidation, i.e., territory reduction in that condition in which it stayed before investigation and extraction of power raw materials.

Extraction and transportation of **natural gas** in comparison with oil represent essentially smaller ecological danger since gas has relative density 0.6 and emitted, quickly dissipates in the atmosphere. Durability of gas pipelines is not rated, but experts consider 30–40 years to be a no-failure service life. A number of actions provides relatively high degree of technological safety: mainly underground lining and placing gas pipelines outside housing estates, and regular and rigid control of all pipeline system elements, their condition. As a result, failure frequency has decreased approximately by 2.5 times for the last 20–25 years.

The essential harm to the nature is done by extraction and preliminary preparation of **solid fuel** (**coal** first): land alienation and landscape destruction when extracted by open quarry, surface subsided by mining, gangue storage problems, etc. When coal is transported, it is almost impossible to avoid such problems as pollution of the grounds adjoining to railways and highways.

Serious damage to the environment does extraction and primary processing of uranium ore:

1. In the production by the mine (approximately 50%) or pit method of uranium ore, it is spread by wind and water over long distances.
2. In the course of processing the ore at the concentrators, a huge amount of waste is produced—the "tails" (more than 500 million tons), which will remain radioactive for millions of years.
3. When processing uranium concentrate into nuclear fuel gaseous and liquid radioactive wastes are formed (fortunately, the radiation dose from them is much less than in the stages of extraction and processing of uranium ore).

Radical improvements of ecological conditions cannot be achieved only by taking inactive measures aimed at environmental protection (various clearing device application) owing to economic restrictions. For instance, unit cost of fume clearing from SO_2 and NO_X increases by several times at the attempt to raise the efficiency of gas cleaner units over—5–85%.

8.3.2   Energy Generation

The biggest environmental problems, as mentioned above, are due to the production of electricity and heat at power plants and boilers that burn fossil fuels, forming "fuel triad"—coal, oil, natural gas, Fig. 8.9.

Power industry development based on coal fuel, simultaneously with the solution of a problem concerning reliable power supply (due to huge world coalfields), aggravates an environmental contamination problem. Annual world emissions produced by power units constitute: carbonic gas—$(2\text{–}3) \times 10^{10}$ t, particulate matters—2.5×10^{10} t, NO_X—1.2×10^9 t, SO_X 1.5×10^8 t, and considerable amount of toxic components. One coal power plant with capacity of 150 MW and average technical characteristics produces more than 1 million tons of greenhouse gases emissions a year. Centralized heat supply provided by means of thermal power plants and nuclear power plants is accompanied by large amount of low-grade heat emissions produced by industrial water cooling systems and by additional fuel expenses on delivery of water heating.

Experience of the advanced countries (first of all Germany, USA, England, Denmark, Japan, South Korea) has proved that modern level of technologies allows reducing ecological impact of coal power stations to a minimum, which is possible only if additional measures are implemented, at the same time making construction and operation of such thermal power stations more complicated and expensive.

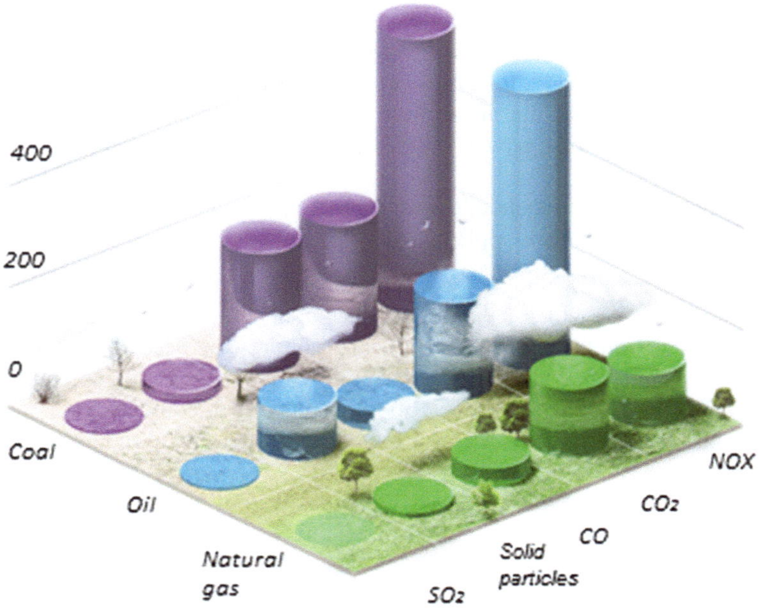

Fig. 8.9 Share of different fuels in overall emissions of 5 pollutants

Table 8.4 Predictable estimation of atmospheric emission reduction, thousands tons per year

Pollutants	Use of CHPP[a] and GTPP[b] for natural gas combustion	Increase in fuel combustion efficiency	Installation of highly effective cleaning units	Introduction of actions for achievement of standard specifications	Total
NO_x	5.5	10.0	11.5	304.6	331.6
SO_2		37.2	165.3	58.2	260.7
Ashes		30.1	295.3	280.0	605.3
Total	1197.6				

[a]Combined heating power plant
[b]Gas turbine power plant

Table 8.4 gives the estimation of atmospheric emission reduction because of implementation of two action groups: technological and organizational.

Nuclear fuel and renewable energy sources produce the so-called fuel "triad" in terms of environmental pollution. It is necessary to bear in mind that nuclear-power engineering, as well as renewable, can be related to "pure" generation if consider operation of power installations only. However, enterprises, which provide their functioning, are the sources of environmental contamination (indirect damage to the environment). After Chernobyl and Fucusima-1 nuclear power plant accidents ecologists are seriously concerned about safe operation of nuclear power plants, storage of the spent fuel and radioactive wastes, treatment of the spent nuclear units, and after New York terrorist attacks in 2001—physical protection of nuclear units.

In technologies of electricity generation from renewable energy sources, there are no processes of oxidation, which could pollute environment; however, they have negative direct and indirect impact on the environment. Among direct impacts are earth alienation, audible noise, bird migrations blocking (wind power), earth surface shadowing (solar power), corrosive solution release (geothermal power), etc.

In order to increase economic and ecological efficiency of conventional fuel usage some approaches are applied.

1. Increase in the parameters of boiler power units with traditional coal burning, and with boilers, which have circulating fluidized bed combustor. It will raise the efficiency up to 53–54% in the nearest future (see Sect. 4.2).
2. Improvement of fuel combustion technology: application of low-toxic burners, boiler transfer to a technology of three-stage fuel combustion, combustion chamber modernization, and introduction of automatic combustion mode management.
3. Special treatment of fuel before its combustion (thin reduction and preliminary coal gasification, coal-water slurry preparation and water black oil emulsion, etc.).
4. Introduction of CHPP and GTPP, providing both high energy and ecological efficiency and reducing specific fuel consumption by 32–37%. (Prevailing part

of TPP and boiler-houses maintained in Russia were constructed at the time when energy resources were cheap and that is why nowadays they do not meet the requirements of cost-effective use of resources and ecology.)

5. Cogeneration—co-production of electric and thermal energy (see Sect. 4.3).
6. Low-grade energy recycling: recycling of reduced steam energy, development of turbo expander technologies, recycling of heat energy produced by technological and household dumps, etc.
7. Wide use of modern highly effective heat exchange devices.

The essential economic and ecological effect provides involvement in energy production of backup fuel (see Sect. 3.2) and renewable energy sources (see Chap. 5).

8.3.3 Energy Transmission and Distribution

Damage caused by transmission and distribution of electric and thermal energy is not so considerable compared with its generation and production. Nevertheless, the transmission of energy over long distances and its distribution among the consumers as well as the early stages of the "life cycle" of energy and power have a negative impact on the three main subsystems: environmental, economic, and social.

1. Environmental and economic subsystem:

   - Violation of the soil-plant complex and relief,
   - Changes in habitat of animals, birds, insects, and their gene pool,
   - A negative influence on the biological processes in the vegetable world,
   - Limiting and changing migration routes of animals and birds,
   - Rejection of valuable agricultural land,
   - Deforestation for the track OTL.

2. Social subsystem:

   - Deterioration of living conditions of population close to power lines (acoustic noise and television and radio interferences),
   - Negative aesthetic impact on the landscape, human settlements, recreation, cultural and natural monuments, etc.,
   - Negative impact of electromagnetic fields of OTL on the human in the conservation zone.

Measures to reduce the negative impact of OTL on the environment are diverse. They can be summarized as follows:

1. Increase the economic interest of grid companies to reduce damage to the environment (fee for the use of land, forest, and other natural resources according to their real value).

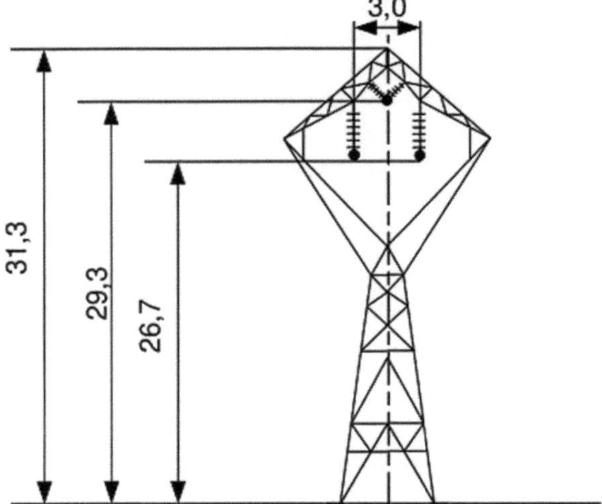

Fig. 8.10 Support
single-circuit compact
220 kV transmission line

2. Use of cable lines in densely populated areas.
3. Regulation of the behavior of population and economic entities in the area of influence of OTL on the grounds of safety.
4. Protection of workers from the adverse effects of electromagnetic fields and high voltage installations
5. Use of technical (technological) measures:

   - Optimization of the design of transmission lines supports (for example, compact transmission lines) and their routes, taking into account environmental effects, Figs. 8.10 and 8.11.
   - Use of the combined power transmission lines—placing on the electricity supports of high and ultra-high voltage transmission wires of lower voltage class,
   - Creation of a transmission plant array of trees and shrubs (preferably fruit and berry species) with plant height slightly exceeding the height of a man.

Failures in pipelines, while transmitting thermal energy, can lead to city landscape destruction (owing to excavations and flooding) and even to human victims.

Fig. 8.11 Support compact double circuit 220 kV transmission line

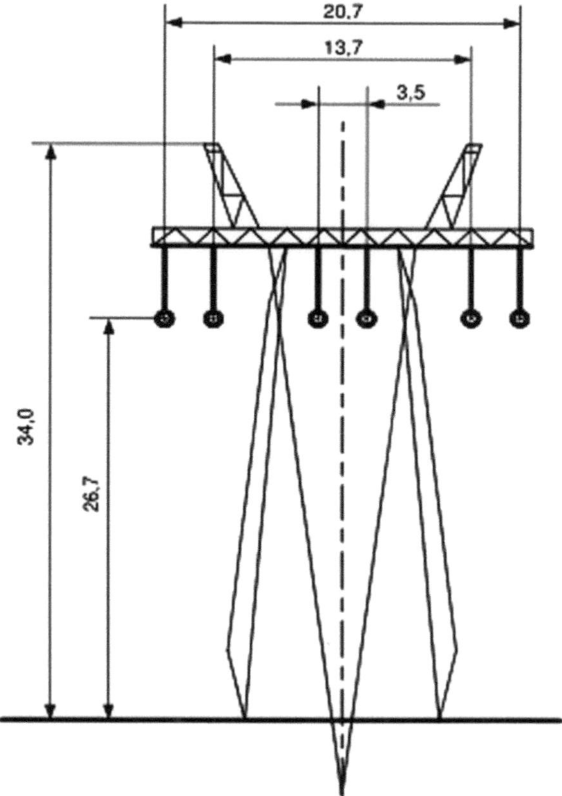

8.4 Look of the International Community of the Environmental Protection Problem

Global environmental changes and local environmental disasters have contributed to the realization to the mid XXth century scale of the problem, especially by the industrialized countries, which are major polluters of the environment. After the Great London Smog (5–9 December 1952), which claimed the lives of up to 12 thousand people, scientists began systematic studies of contamination of all geospheres, regulation of emissions, discharges, and waste disposal, improvement of cleaning systems, introduction of low-waste technologies, the formation of regulatory framework environment protection, and management of natural resources.

The attention to the protection of the environment was moved to the level of parliaments, governments and international organizations formation of international environmental organizations (including the UN Commission on Illustration quality thermal insulation of pipes of thermal energy transport systems. Environment and

Development), and intensive improvement of the relevant legislation. In 1968, a group of scientists and businesspersons from different countries organized the Roman Club—the international non-governmental organization aimed at investigating global problems and methods for their solution. The first Club report written by Donella and Denis Meadows, Jorgen Randes, and B.B. Berens entitled "Limits of Growth" was published in 1972. This report attracted attention of politicians and scientists all over the world; it stated that the uncontrollable growth of population, ruthless exploitation of natural resources, and pollution of the environment brought the threat to the future of humanity. Some people have perceived "Limits of Growth" as a prediction of the close end of the world.

More than 30 years later the authors of the first report have corrected their computer model and published in 1992 one more report entitled "Outside: Global Catastrophe or Sustainable Future?" and book entitled "The Limits to Growth: The 30-year Update" [1]. A new Roman Club Report entitled "Factor 5: Transforming the Global Economy through 80% Increase in Resource Productivity, [2], in which new solutions of old problems trapping humanity on the way to sustainable development were suggested. (According to opinion the most of scientists and experts, this book is one of the best publications devoted to these problems.)

Is the correct style of your life? How to life further? These are the main questions the authors of this book try to answer. This book deals with economics, technology, natural resources, and free market. Since the time of industrial revolution, the term progress taken to mean an increase in *the labor productivity*. A new approach to the progress is suggested in the book, which makes an increase in *the **productivity of resources*** the corner stone of the progress. As the authors stated, we could live twice better and at the same time, spent half as many resources as at present. This is necessary for the sustainable future development of humanity. A solution is to increase the efficiency of utilization of energy, water, fuel, materials, fertile soil, etc., often without extra costs and even with profit. As the book demonstrates, the majority of engineering solutions of our problems have already been obtained; they must be used right now.

We reasoned in due time about energy-saving policy the quintessence of which can be expressed in the well-known notice on walls of institutions: "Before leaving, switch off the light!" Thus, the productive utilization of resources is not new. What is new is the huge number of unrealized capabilities we have. The authors give tens of examples from everyday life. They not only provide recommendations (sometimes rather simple) but also realize many of them in practice. The book abounds with practical examples of technologies that permit more efficient utilization global resources. It can become a reference book for those who want to understand how the technology serves purposes of sustainable development and environmental protection. Unfortunately, in our everyday life we face tens of counterexamples from taps through which seas of precious pure water escape up to heating mains in cities that are rebuilt every 3–4 years and nevertheless, their thermal insulation is such that in winter snow melts over them, Fig. 8.12.

Fig. 8.12 Illustration of poor quality thermal insulation of pipes of transport systems of thermal energy

There have been numerous attempts to explain how to organize the markets and to reorganize the system of taxation so that the welfare of people could grow without increase of the consumption of resources [2].

There are several different approaches to the problem and focuses on whole sectors of the economy (construction and operation of buildings, production of steel and cement, agriculture, transport), which are in the world consume the greatest amount of energy, water and mineral resources and create the greatest amount of greenhouse gases. It is shown that in all the sectors actually achieve the fivefold increase resource efficiency.

By the early 90s of the XXth century the basis of the concept of sustainable development well known today has been formed. The concept was adopted in June 1992 at a conference in Rio de Janeiro and confirmed 10 years later in Johannesburg.

In order to unite the efforts of countries to prevent dangerous climate change and to stabilize greenhouse gas concentrations in the atmosphere at a relatively safe level, it was adopted in 1994 and entered into force on United Nations Framework Convention on Climate Change (UNFCCC). Member countries of the Convention concluded that greenhouse gas emissions should be reduced on a mandatory basis; controversy is only safe level of emissions.

The global nature of the problem was adopted at the Third Conference of Parties of the UNFCCC in 1997 in Kyoto (a Japanese city) by a special document, fixing quantitative commitments by developed countries and countries with economies in transition limit and reduce greenhouse gases entering the atmosphere, as well as mechanisms of implementation of these commitments by 2012.

The document is called Kyoto Protocol. According to this document, countries have committed themselves to control and limitation of polluting emissions into the atmosphere.

Among the countries that signed the KP, there were those for whom its implementation carries obvious benefits: economic, social, political, and environmental. However, there were those for whom the manual is not very profitable in the usual "mundane" sense, primarily because of its "inhibitory" effect on the pace of industrial development.

One of the main barriers to more efficient utilization of resources is a contradiction between the developed and developing countries. For the letter, saving of resources and careful attitude to nature often retreat in the face of momentary problems of eradicating poverty they try to solve by choosing the western way of the development, alas, not free of many errors.

The results obtained for the years of the KP execution (1997–2012) lead to the conclusion that the KP is only a pilot phase of a global, long-term process aimed at preventing human impact on climate, and the environment and that the KP mechanisms are not fully debugged.

The international community will have to overcome the contradictions on the measures to limit greenhouse gas emissions between developed and developing countries. (The latter insist on allocating those funds for environmental protection as a compensation for the pollution of the planet by developed countries in the past.)

The consequence of these contradictions is modest progress in the development and coordination of post-Kyoto agreement at international forums in 2009—Copenhagen (Denmark), 2011—Durban (South Africa), and 2012—Doha (Qatar). Their main result can be considered as a development scenario for further actions to limit GHG emissions. In it, all countries are divided into three groups according to the level of socio-economic development, which defined their obligations to the world community:

1. Developed countries should cut greenhouse gas emissions by 2020 by 60–80%.
2. Major developing countries must start to reduce GHG emissions within 2020–2030 years.
3. African countries in the XXIst century do not take the commitment to reduce GHG emissions.

The latest summit dedicated to environment protection (limitation of the greenhouse gas emission) was held in Paris (from 30 November to 12 December 2015.)

Representatives of 196 countries and leaders of 150 countries attended the event. Some of them took advantage of this podium to emphasize the critical importance of developing a common approach to solving this problem. Because of nearly two weeks of discussion, it was concluded the first ever global and binding agreement to reduce greenhouse gas emissions was developed. Governments agreed mandatory, following rigid and transparent rules, to ensure that all countries fulfill their commitments. All countries agreed to pursue decline of global warming to below 2 °C.

This made a significant contribution to efforts to achieve the declared objectives in the UN aim—"to save succeeding generations from the disasters." The agreement resolved the controversy over the financial side to solve the problem between developing and developed countries—in the period before 2020 developing countries will receive from other countries 100 billion dollars annually. UN Secretary General Pan Gi Mun called the agreement "health insurance for the entire planet."

For many developing countries, the revolution in the efficiency can give the unique opportunity of prosperity for a sufficiently short time period. However, as was discussed in the World Ecological Forum held in Rio-de-Janeiro in 1992, the new way of thinking was acceptable not for all people.

In conclusion, of this chapter, we consider one more extremely efficient way of preservation of the Earth as a cradle and a unique suitable human habitat. This way of conscious self-restraint and change of the lifestyle is extremely difficult in use.

Life in developed countries differs by an extraordinary high rate of consumption of natural resources. In particular, these countries burn more than halve the fossil fuel whereas their population is only 20% of the Earth's population. Such rate of consumption of resources is determined by the life standards that have already been reached by the majority of population of these countries and for which the remaining countries strive: a separate rather large house and one or better two cars per family, an opportunity to travel, etc. These standards finally determine the vital needs of the society in black and non-ferrous metals, cement, plastic, wood, etc. and through the necessity to extract, to treat, and to transport all this demands in energy.

A question arises: can the population of developed countries anyhow change their life standards? Do the old life standards (large house, two cars, etc.) remain so much attractive if we take into account that following this lifestyle we will soon convert our Earth into a desert?

Is there any acceptable alternative to this lifestyle? Freedom of moving can be met with the developed system of public transport, which is certainly more economical than personal cars. In addition, as to a large house, may be the life in small houses (or even in blocks of flats) surrounded by almost wild nature is more attractive? Such questions evoke a smile now, but it is feared that they will be unavoidably posed soon in the course of world evolution.

By the way, the transition to this style of life for residents of developed countries will not necessarily be accompanied by their feelings of loss. Most likely, reasonable self-restraint in combination with high tech will only increase the quality of life in these countries, though due to new components in comparison with already existing ones.

It should be emphasized that ecological actions are not attractive to investors and consequently can be carried out, basically, at the state level for the account of rationing, stimulation, and budgetary appropriations. For environmental protection, it is necessary to include in ecological programs actions of all levels aimed at energy saving, as they are more investment-attractive.

Periodic meetings of the world countries leaders in which sustainable development of the human civilization problems are discussed, gives hope that the collective efforts enable humanity to cope with this threat.

Questions and Tasks to Chapter 8

1. Describe the changes in the energy balance from the beginning of the XXth century to the present day.
2. Compare the power of all worlds' power plants with a capacity of some natural phenomena.
3. Nature of the "greenhouse effect" and its most dire consequences.
4. Describe the changes in concentration of CO_2 in the atmosphere in the course of the XXth century and 1.5 decades of the XXIst century.
5. How can describe the atmosphere temperature changes for last century?
6. Give the quantitative characteristics to the changes of see level and the Earth surface temperature.
7. What alternative climate changes hypothesis do you know?
8. What is generally meant by "sustainable human development"?
9. What factors influence on the extent of threat to the environment?
10. Name the main measures for protection of the humanity from threats caused by anthropogenic influence on the environment:

    - primary energy consumers,
    - producers of the electrical and thermal energy,
    - consumers of the electrical and thermal energy.

11. What are the effects providing effective power production and consumption?
12. What measures are undertaken to avoid dangerous pollution of the Earth?
13. Why the use of mineral oil for electric power production is criticized?
14. What changes have taken place in the energy consumption after 1973–1974?
15. Main negative factors influencing on the environment in the process of primary energy resource extraction:

    - coal,
    - oil and natural gas,
    - Uranium ore.

16. What are the main ways to reduce the impact of resource FEC sectors on the environment?
17. Ways to reduce negative impacts on the environment in the process of electrical energy production.
18. Ways to reduce negative impacts on the environment in processes of electrical energy transportation and distribution.
19. What are the main factors that impede to overcome the 7th ecological crisis?
20. Directions of action of the international community to protect the environment.
21. What are the main results of summit on environmental protection (30 Nov.–22 Dec. 2015, Paris)?

23. Why the leaders of developing countries do not want to accept limitation of CO^2-emission as one of the main tasks of the governments?
24. Why do the governments of all developed countries stimulate energy saving?

References

1. Dunsheath P. A history of electrical power engineering. MIT Press Classic; 1969.
2. Weizsaecker E, Hargroves K, Smith M, at al. Factor 5: transforming the global economy through 80% increase in resource productivity. London, UK: Earthscan; 2009.

Chapter 9
Unconventional (Alternative) Methods of Electric Energy Production

Most research associated with alternative methods of power generation have concentrated on relatively high efficiency systems, as this makes commercial realization more readily achievable. To eliminate a threat of the forthcoming energy shortage, we are pinning our hopes on mastering of new power sources and on an increase in the power of already mastered methods of direct thermal, nuclear, and chemical energy conversion into electric energy. These methods will radically increase the efficiency of utilization of materials involved in power generation processes and significantly increase the amount of planetary resources accessible for practical utilization.

In these sources, 2 types of effects are used:

- Physical ones that are manifested in different forms including synthesis of light elements, photoelectric effect, thermo-electricity, etc. (fusion reactors, thermionic generators, photoelectric batteries, thermionic-emission generators, etc.)
- Chemical ones that are manifested in an energy release in redox reactions of chemical reactants (galvanic cells, accumulators, etc.).

Controllable thermonuclear fusion (tokamak and inertial (pulsed) thermonuclear fusion), electrochemical generators (hydrogen power engineering based on fuel cells) together with thermoelectric, thermal emission, and radioisotope power sources belong to unconventional methods of electric energy production. At present, only the first two non-traditional method of producing electrical energy meet the requirements of large-scale electrical engineering. However, their large-scale use in the energy sector can be expected only after a few decades. Other methods mentioned above have already been well utilized for power supply of low-power consumers. The output of a large power is unlikely and so in this textbook, they are not considered.

The direct power generation has found the widest acceptance in autonomous low-power sources for which reliability, compactness, and convenience in service, and low weight rather than the economic efficiency are of great importance. They are mainly used for data acquisition in almost inaccessible localities of the Earth, in

© Springer International Publishing AG 2018
V.Y. Ushakov, *Electrical Power Engineering*, Green Energy and Technology,
https://doi.org/10.1007/978-3-319-62301-6_9

space, and on board airplanes, vessels, space vehicles, etc. As a rule, conventional generators, based on combustion of chemical fuel and large-scale method of thermal energy conversion into electric energy, require regular service; therefore, they are unsuitable for the considered purposes.

The aggregate installed capacity of millions of autonomous electric energy sources, despite their moderate sizes, exceeds the capacity of all stationary power plants taken together.

9.1   Thermonuclear Power Engineering (Controlled Thermonuclear Fusion)

One of the possible ways to satisfy the world's population energy needs is mastering of controlled fusion of light elements—controlled thermonuclear fusion (CTF)—since virtually inexhaustible power source will be obtained.

The main portion of energy of stars and the Sun, as was proved by Hans Bethe in 1939, is released in the process of synthesis of light elements. If one succeeded in creating conditions for a controlled thermonuclear reaction of light element fusion (deuterium, tritium) to occur on the Earth, this, figuratively speaking, will mean the appearance of the artificial small suns on our planet capable to satisfy energy needs of many generations.

For practical use in the energy sector today is considered basically three reactions:

$$D + T \rightarrow n + {}^{4}He \tag{9.1}$$

$$^{3}He + D \rightarrow p + {}^{4}He \tag{9.2}$$

$$D + D \begin{cases} \nearrow & n + {}^{3}He \\ \searrow & p + T \end{cases} \tag{9.3}$$

D and T   isotopes of hydrogen: deuterium and tritium,
n and p   the neutrons and protons, respectively,
3H, 4H   three and four helium nuclei, i.e., alpha particles

More clearly, the thermonuclear reaction of fusion of hydrogen isotopes—deuterium and tritium is shown in Fig. 9.1.

The reactions (9.1) and (9.2) call priority interest. The first—thanks to the largest number of energy released, the second—because of a simple solution to the problem of "fuel" and the lack of induced radiation.

It is very important that the fuel for fusion reactors is easily extractable. Deuterium occurs naturally in water: one of 6700 hydrogen atoms has a deuterium core. Tritium is less widespread. It is radioactive and has a half-life of 12.3 year, so

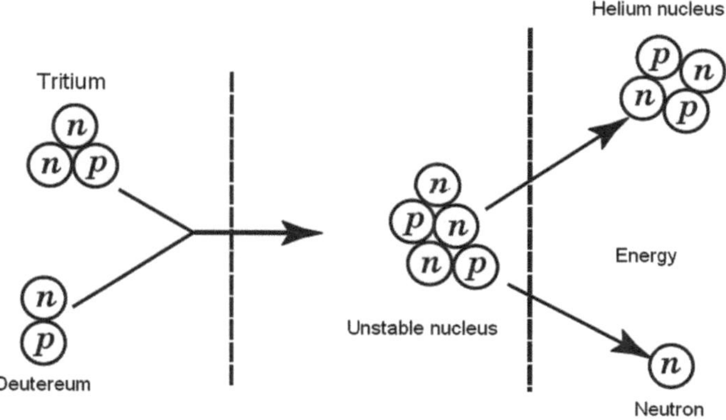

Fig. 9.1 Thermonuclear reaction of fusion of hydrogen isotopes deuterium and tritium

that it does not occur naturally in large amounts. However, tritium can be produced from the lithium or lithium salts in the blanket of the vacuum chamber. Neutrons flying out of the plasma by reaction with lithium give the most part of energy for heating lithium and production of tritium.

Reaction (9.2) is attractive, the first of all, because Helium-3 (3He) in large quantities (approximately 500 million tons) is available on the Moon, wherefrom it can be delivered to the Earth. Several countries have announced plans for mining minerals on the Moon, especially helium-3: United States, Russia, China, India, Japan, the European Space Agency, and others. The US has already developed an appropriate working draft. NASA management proposes to make the project of an international lunar base, the type of the International Space Station. At that time the efforts of physicists focused on the technological embodiment of reaction (9.1), i.e., on the synthesis of deuterium and tritium, releasing energy in the quantities of interest to the "big" power engineering. Concerns about fuel for controlled thermonuclear fusion (CTF) commercial scale are somewhat premature, taking into account the complexity of problems in its implementation, on which physicists have been working for about 60 years.

Despite many-year studies of controlled fusion, construction of an industrial reactor is still many years in the future. To overcome the natural electrical repulsion, the nuclei must have significant energy. The temperature of a deuterium-tritium mixture must reach, at least, 50 million degrees (for comparison, the temperature in the center of the Sun is about 15 million degrees). Measured in electron volts (eV), this temperature is 4500 eV. At this temperature, the electrons are turned loose from nuclei (in fact, the energy of only 13.56 eV is necessary for hydrogen atom ionization). The deuterium-tritium mixture in this case represents a plasma—an electrically neutral gas consisting of positively charged nuclei and negatively charged electrons.

The maintenance of high plasma temperature is one of the main problems of thermonuclear research. The plasma loses energy as a result, of several processes.

For example, charged particles in plasma radiate electromagnetic energy in the process of collisions with each other. In thermonuclear reactions, a huge number of fast neutrons, which then easily escape from plasma, are produced. The radiation, heat conduction, and turbulent convection of particles in the plasma are among the processes resulting in plasma cooling and a decrease in its temperature.

The "plasma fire" can be continuously supported by energy supply from the outside with the help of radio frequency waves or high-energy beams of neutral particles. However, there is an efficient self-sustaining source of additional heat—fast α-particles generated in plasma. These helium nuclei are the "ash" of thermonuclear reactions. They are generated with energy of about 3.5 million eV and are easily confined with a magnetic field because they have a double positive charge. In the process of collisions with plasma particles, α-particles give up energy to them in the form of heat. Up to now, the researchers have failed to generate sufficient number of high-energy α-particles in order that to compensate completely for thermal energy losses.

The researchers call the total average time during which the plasma looses heat the time of energy confinement or the energy time τ. The product of τ and plasma density n characterizes the capability of plasma to conserve heat and is called the parameter of heat conservation quality. In order that thermonuclear reactions were self-sustained and gave useful power, the product $n\tau$ must be greater than 2×10^{20}, where time is in seconds and density is in the number of particles per cubic meter, at the temperature $T = 10{,}000$ eV (about 100 million degrees). Thus, the thermonuclear research is aimed at obtaining the product of three values: n, τ and T of about 2×10^{24} s eV/m^3 (Lowson criterion).

9.1.1 Tokamak

Thermonuclear systems called *tokamaks* are best suited now for this purpose. Russian scientists A.D. Sakharov and I.E. Tamm invented this system in early 1950. The name of this system is an abbreviation of Russian words "the toroidal chamber with a magnetic field." The principles underlying the system operation with a magnetic field are shown in Fig. 9.2.

First the plasma is ignited in a vacuum chamber shaped as a torus. The system of electromagnets arranged outside of the chamber creates a toroidal magnetic field directed along the torus axis. The field acts as a hose, which maintains the pressure inside the plasma and prevents its contact with chamber walls.

Other system of electromagnets arranged in the center of the torus (in the hole) is used to induce an electric current running in the toroidal direction in the plasma. This current heats the plasma to a temperature of about 1000 eV. The plasma current creates its own magnetic field encompassing the toroid. This field prevents a drift of plasma particles outside of the region of magnetic confinement. Finally, the external conductors of the poloidal field magnet generate a vertical magnetic field

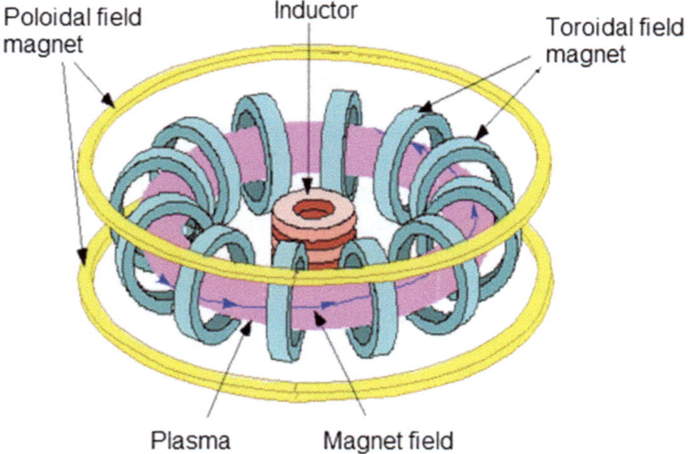

Fig. 9.2   Three electromagnet systems of a tokamak

confining the plasma column from upward and downward drifts as well as from drifts to the left and to the right inside of the chamber.

Up to the middle 60s, the tokamak concept had been systematically investigated only in the former Soviet Union. L.A. Artsimovich with colleague from the I.V. Kurchatov Institute of Atomic Energy have managed to gain knowledge which convinced physicists from other countries that the period of energy confinement and the temperature of plasma in tokamaks are promising for CTF implementation. Their achievements gave impetus to the research of tokamaks in a number of the world's leading countries.

It is important that research on controlled thermonuclear fusion is international in character. When visiting England with our governmental delegacy in 1956, Academician I.V. Kurchatov told about the most interesting results obtained in this area in our country. Later other countries followed the example of the Soviet Union. It turned out that, working independently and secretly, the scientists of different countries came to the same ideas of implementation of controlled thermonuclear fusion. Since then these studies has become open with wide exchange by the results and discussion of perspective ideas.

As a result, the tokamak design has been significantly improved. In the middle 70s, a temperature of 3000 eV and a parameter of confinement quality of about 10^{18} s/m^3 were realized in tokamaks. At present, a plasma temperature of 30,000 eV and a confinement quality parameter of 2×10^{19} s/m^3 are realized in the most powerful experimental systems of this type including JET (Joint European Torus), JT-60 (Japan), and TFTR (Tokamak Fusion Test Reactor being an experimental thermonuclear reactor) tokamaks and DIII-D system (USA). In 1970–1990, the product of density, confinement time, and temperature increased more than 100 times.

Qualitatively new stage of collaboration in solving the CTF problem started in 1985, when the leaders of the USSR and USA (M.S. Gorbachev and R. Reagan) appeal for international collaboration in mastering thermonuclear energy for the benefit of humanity when they met in Geneva. In respond to this appeal, the engineers and scientists participating in four leading programs of research on thermonuclear fusion carried out in the European Community Countries, Japan, the USSR, and the USA decided to start in 1987 joint design of the experimental thermonuclear system. They called it International Thermonuclear Experimental Reactor (ITER).

Despite a definite progress in experiments on tokamaks, some problems still remain to be solved. The researchers have not yet elucidated the fundamental nature of turbulent heat and particle transfer crosswise the force lines of a magnetic field—the process that reduces the plasma temperature. The knowledge of physics of combustion and sustenance of thermonuclear combustion is also incomplete. These problems will be investigated on the ITER. The ITER Project is aimed at the provision of conditions of ignition and long-term thermonuclear combustion, typical of an actual thermonuclear reactor, as well as at testing and demonstrating technologies of practical application of controlled fusion, Fig. 9.3.

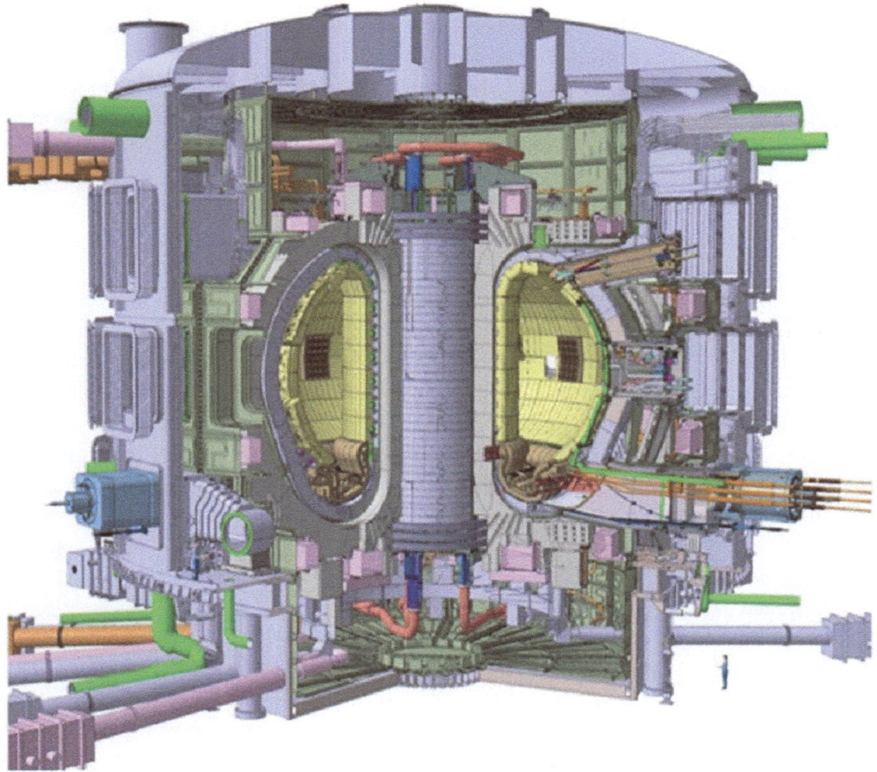

Fig. 9.3 Cross sectional view of ITER

The experimental reactor under construction will be largest of tokamaks ever constructed: its altitude is 30 m and its diameter is 30 m. The plasma volume in the reactor (850 m^3) is also very large; the current in the plasma is 15 MA. A designed capacity of the thermonuclear reactor will be 500 MW for 400 s. This time will then be increased to 300 s to enable the researchers to study for the first time the physics of thermonuclear combustion in the plasma.

According to the expert estimates, the cost of the ITER design and construction will be about 7.5 billion dollars. More than 200 institutions are involved in the project implementation only from the Russian party.

Upon completion of the reactor, the ITER program will consist of two main stages. The first stage, called physical one, will last for 6–8 years. In this time, the researchers will try to provide conditions for ignition and long-term combustion typical of high-power thermonuclear reactor. After putting into operation and full testing, research will be focused on the sustenance of the stable plasma and the provision of conditions for deuterium-tritium fusion. Effects of plasma heating by α-particles, dynamics, and control of plasma combustion will also be studied together with diffusion and elimination of helium once the helium nuclei have given up all their energy to the plasma. The second stage will be a long-standing program of solving technical and engineering problems. Many technologies will be demonstrated already at the physical stage. The most important among these technologies are the operation of superconducting electromagnets, systems of plasma heating and maintenance of the current, systems for fuel feeding and deleting "ash," instruments for remote servicing and maintenance, and external supporting systems. In the technological stage, the integral characteristics and reliability of the equipment will be tested as well as alternative materials and designs.

The design and engineering work must result in the creation of the ITER reactor with a design capacity of 1000 MW generated in the process of fusion of deuterium and tritium. This will be a considerable achievement. The expected thermonuclear yield will be about three orders of magnitude greater than that already produced by the JET being the most powerful thermonuclear reactor now.

The capability to generate thermonuclear power 1000 times greater than that produced by the existing experimental systems will make the ITER the penultimate stage for the practical implementation of controlled thermonuclear fusion.

After repeated corrections due to unforeseen technical and organizational problems, the coordinating center has approved the following schedule of the project:

- 2010–2019—the construction of ITER,
- 2020—a pilot launch,
- 2020–2037—experimenting, testing of all ITER systems,
- 2050—creation of the demonstrational thermonuclear power plant (project "DEMO"). Its most likely power circuit shown in Fig. 9.4.

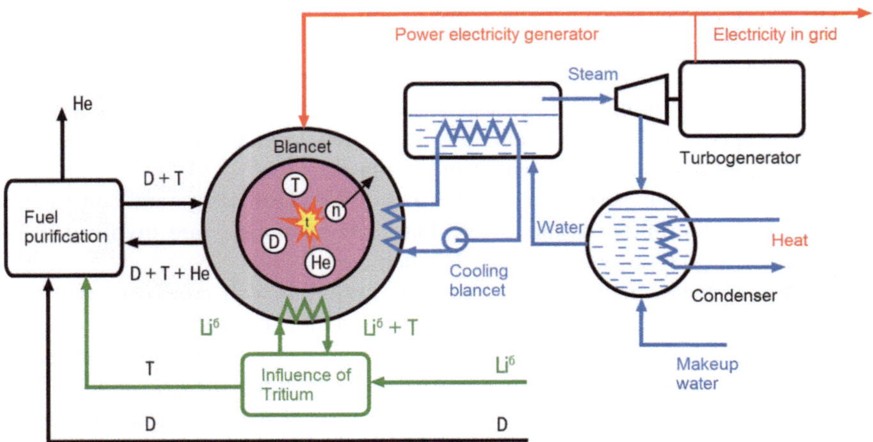

Fig. 9.4 Scheme of the main technological fusion power reactor operating on a mixture of deuterium and tritium

Now the USA, Russia, EU, Japan, China, South Korea and India are involved in the project ITER. Brazil, Kazakhstan, Canada, Mexico also express the desire to take part in the project.

9.1.2 Inertial (Pulsed) Thermonuclear Fusion

Almost simultaneously with the development of tokamaks, another scheme of the thermonuclear reaction of fusion was launched—the creation of pulsed systems using the effect of pinching (compression) of the plasma column containing thermonuclear fuel or compression of deuterium-tritium target with the help of the compression effects of powerful ion beams or laser pulses.

The compression fusion reactors with gaseous or solid thermonuclear fuel in the initial state enclosed in a spherical shell is exposed to ion beams or high-power pulses of electromagnetic radiation (laser or X-ray) from external sources. Under the action of radiation, shell material evaporates and creates a reactive force that can compress and heat membrane and fuel to the densities and temperatures under which the Lawson criterion is fulfilled, Fig. 9.5.

This reaction is called "Inertial thermonuclear fusion (ITF)". Developed countries (the USA, the USSR, GB, France et al.) have started the realization this idea in the 60s the last century.

This direction, which is in many respects alternative to the first one, instead of the confinement of unstable plasmoids, is oriented at the creation of conditions (the plasma density) under which the main portion of thermonuclear fuel would be depleted faster than these plasmoids are separated. The time-dependent parameters

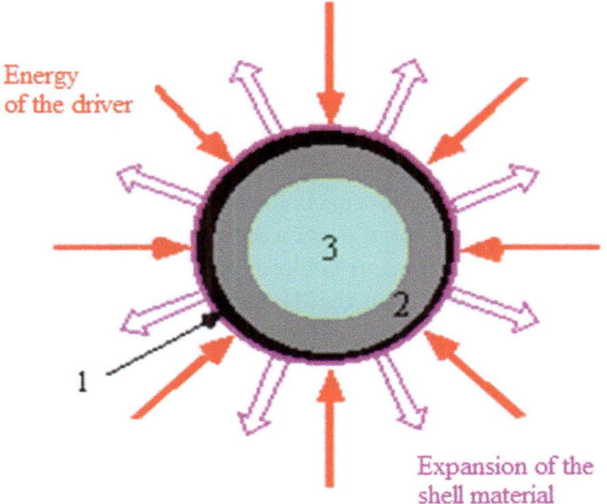

Fig. 9.5 Principle of implementation of inertial thermonuclear fusion: *1* hollow shell, *2* DT layer frozen mixture, *3* DT low-density gas

of this process are determined by the inertia of the fuel mixture. In the creation of a pulse thermonuclear reactor, the problem of plasma confinement is transformed into the problem of plasma heating for a very short time. At present, the creation of pulse thermonuclear reactors is at the stage of substantiation of conceptual projects.

In order to achieve the desired results in the field of the laser pulse and ion beam ITF, technical problems of enormous complexity must be solved. It is necessary to ensure that:

- high-power lasers and accelerators (with a peak power at 10^{15} W) operated in high-frequency mode (10–100 shots per second);
- their efficiency is increased from current 0.3 to 10–15% (at least);
- compression of the target by means of laser pulses or ions beams is uniform;
- long-term resistance to shock loads of materials at high temperatures must be provided (in explosion of only one target energy is released of several tens of kWh), and others.

The application of pulsed thermonuclear reactors for the construction of power plants will also face economic challenges: extremely high cost of construction and operation of stations and, as a consequence, high price of electricity produced. Nevertheless, nevertheless, it is the direction in the development of fusion continues to develop in England, France, America, Japan, and Russia.

The efforts of scientists have resulted in the fact that now pulse thermonuclear fusion starts to compete by a number of parameters with the more conventional magnetic plasma confinement.

The development of thermonuclear reactors operating in the short pulse mode upon exposure to laser or ion beams depends largely on the creation of high-efficiency lasers and high-current accelerators. Nowadays their efficiency is still very low.

The complex problem is the development of a system of utilization of thermonuclear energy that would be capable to escape harm under conditions of repeated explosions of deuterium-tritium tablets upon exposure to laser or ion beams. To obtain an acceptable energy output, rather high frequency of explosion repetition is required, similar to that of ignition of a fuel mixture in an internal combustion engine.

Assessing the prospects for thermonuclear power engineering, we can say that it is promising given if it not were comparing with coal or gas power engineering. By physical principles, it is closer to nuclear power engineering. In this case, we can expect benefits from the thermonuclear power engineering.

1. Calorific value of the thermonuclear fuel is significantly higher than that of nuclear—in synthesis 1 g of D-T mixture, is 8 times more energy is released than in fusion of 1 g of uranium (10–20 million times more than by burning of 1 g of fossil fuel).
2. Thermonuclear power plant is approximately two orders of magnitude safer, then nuclear power plant. The fusion reactors are environmentally safe. The random start of the reactor is impossible, since the amounts of deuterium and tritium in the unit in any given moment are very small. During uncontrollable combustion, all fuel will be fast spent, and the process will be terminated. In addition, only not radioactive fast neutrons and α-particles (helium nuclei) are created due to fusion of deuterium and tritium nuclei. The main problems of irradiation are due to the secondary processes. The high-energy neutrons can cause the transmutation of nuclei in the reactor materials and components, and they can become radioactive. However, investigations have demonstrated that the proper choice of structural materials will allow such induced activity to be kept on a very low level.

The above materials explain the reasons for caution in forecasting the timing of the creation of fusion power plants and even the possibility of solving this problem in the near future [1].

9.2   Hydrogen Power Engineering (Based on Fuel Cells)

Hydrogen energy technologies in most EU countries, the USA, Japan, as well as in Russia, are included in the list of critical technologies, the level of which depends on the economic well-being and security of the country. The results of research in recent years, escalating energy and environmental issues determine the main directions of development of a new market of hydrogen technologies and the use of hydrogen as a clean energy source in the near future.

A common term "hydrogen economy" refers to the economy based on the complete replacement of non-renewable sources; hydrogen fuel energy can reduce emissions of greenhouse gas and harmful substances into the environment. Global demand for hydrogen to 2050 will be 200–300 million tons, and by 2100–800–1000 million tons.

Hydrogen as an energy source is characterized by the following properties:

- Hydrogen stocks are practically non-exhaustible, but in free state it is virtually absent in nature;
- Hydrogen is an universal energy source; it can be used as fuel for electricity generation in the working cycles of different types and as an energy carrier, it is suitable for the transportation in gas, liquid, and bound states;
- Hydrogen can be used for energy storage.

Among other types of fossil fuels, hydrogen has the greatest calorific value per unit weight and the least negative impact on the environment. The energy content of 1 g of hydrogen is equivalent to the energy content of about 3 g of gasoline. When using hydrogen in fuel cells due to their high efficiency (1.5–3 times higher than that of the internal combustion engine), hydrogen fuel efficiency is 4–10 times higher.

When switching to hydrogen fuel, new technical problems inevitable—after receiving it must be compressed, liquefied or saturated by special materials (metal hydrides, chemical hydrides, carbon-based structures) to accumulate it in large amounts, and then transported by care or marine vehicles and stored in storage systems to deliver to user. The life cycle of hydrogen is complete after its consumption.

The main methods producing of hydrogen today is the steam reforming of hydrocarbons (first of all methane) and electrolytic decomposition of water. It is planned to produce hydrogen by thermalizes of water at temperatures higher than 2000 °C achievable in high-temperature nuclear reactors and in the focus of sun-rays. There are suppositions on possibilities of its extraction as a fossil raw material or using step photo catalysis in a set of chemical reactions characteristic of living plants.

Two ways of hydrogen utilization are used today:

- Direct conversion of hydrogen into electricity with help of fuel cells, which can be used simultaneously as electric energy storage devices.
- Burning of hydrogen as the ordinary fuel in some critical technologies:

  - To increase the activity of coal fuel in the powerful boilers of TPP (combustion of hydrogen with the steam can significantly improve the parameters of the steam and consequently achieve a higher efficiency of the steam turbine (55%);
  - To boost operation of turbine unit in the thermal and nuclear power plants during peak load in the electrical network;
  - To overheat steam in order to increase the efficiency of turbine generators (by 1–7.5%) in operation of TPP and NPP.

The fuel cell (FC) is the device, which converts the energy of a chemical reaction directly into electrical energy. The process here involves the oxidation of an

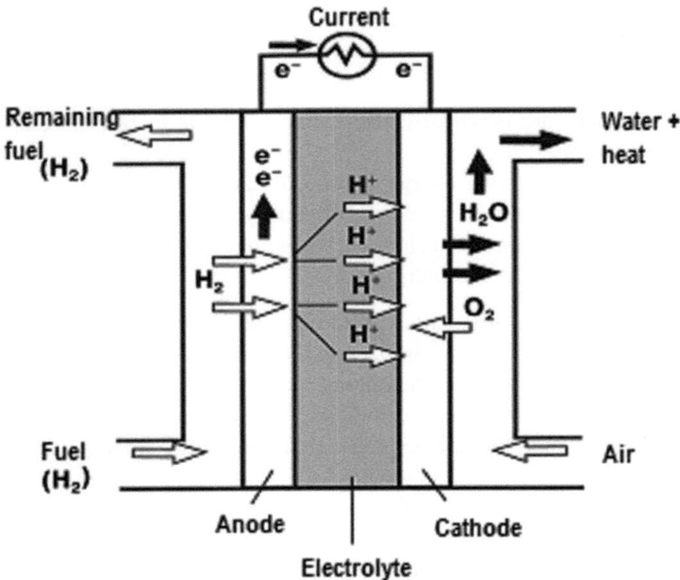

Fig. 9.6 Schematic of an oxygen-hydrogen fuel cell

external fuel, which is normally a hydrogen-rich gas, and an oxidant, which is usually atmospheric oxygen, Fig. 9.6.

The oxygen-hydrogen cells are most developed. The cell can operate at room temperature and atmospheric pressure, but the efficiency of electrochemical process significantly increased with temperature and pressure.

The electrodes in the fuel cell are porous. The positive hydrogen ions are transferred through the electrolyte to the anode. The remaining electrons produce a negative potential and are displaced to the cathode in the external circuit. The oxygen atoms located on the cathode attach electrons and form negative ions; the latter are combed with the hydrogen atoms from water and then are transferred into the solution in the form of hydroxyl ions OH^-. The hydroxyl ions, combining with hydrogen ions, form water molecules. Thus, with continuous supply of hydrogen and oxygen, the continuous reaction of fuel oxidation by ions takes place with the simultaneous flow of current in the external circuit. Since the voltage on the cell leads is low (of the order of 1 V), the cells are connected in series to form a battery. The efficiency of fuel cells is very high. Theoretically, it is close to 100%, and in practice, it is 60–80%. The generation efficiency is much higher in a FC cell at scales even up to several megawatts, it has a higher power density, lower vibration characteristics, and reduced emission of pollutants. (In 1893, German physicist and chemist W. Nernst calculated, that the theoretical efficiency of electrochemical process of coal chemical energy conversion into the electric one is 99.75%.)

Utilization of hydrogen as the fuel leads to high total expense of operation of the fuel cells; that is why the opportunity of using other fuels, primarily natural and gases from special enterprises, is explored. As is will known, gas is a relatively cheap fuel. Unfortunately, the rate of gas oxidation reaction is high enough only at high temperatures (800–1200 K), which gives no possibility of using aqueous alkali solutions as electrolytes. In this case, solid electrolytes with ionic conductivity must be used.

Another complication is that many fuel cells cannot operate effectively using a raw hydrocarbon fuel; most types require a reformer, which converts hydrocarbon into a hydrogen-rich gas suitable for passing directly into the fuel cell. There are many types of FC. The two main types of FC being developed at present are the Solid Polymer Electrolyte Fuel Cell (PEFC) and the Solid Oxide Fuel Cell (SOFC). The type of electrolyte used therein distinguishes these two main classes. This in turn determines their operating temperature.

The PEFC is based upon a proton conducting polyfluorosulphonic membrane with finely dispersed platinum electrodes upon a porous carbon matrix. Machined carbon or steel plates are utilized as bipolar plates to form series stacks. The PEFC operates at temperatures below 100 °C. It is a potentially clean method of generating electricity, which is noiseless, robust and efficient. Some PEFCs are commercially available for demonstration or early application at 1–5 kW for combined heat and power production and up to 25 kW for transport. Expected lifetime is still only a few thousand hours for the stack, but system lifetime is expected eventually to be in the range of 8–20 years.

The SOFC typically comprises a nickel zirconia cermet anode, yttrium stabilized zirconium electrolyte, lanthanum strontium manganite cathode, and a chrome-based alloy interconnect. The SOFC operates at much higher temperatures than the PEFC, in the region of 850–1000 °C. It has high electrical efficiency together with high-grade exhaust heat. Therefore, important applications for the SOFC are in the combined heat and power production and in other applications where a significant heat load is present.

Advantages of the SOFC are very high electrical efficiency, the ability to utilize conventional fuels such as natural gas with limited processing. Typical power ratings are in the range 150–250 kW and life expectancy is 5–20 years, with adequate maintenance. The first demonstration systems in the range 1–2 kW have recently become commercially available, normally in combined heat and power production with about 10 kW thermal output.

At present, the work is underway on the creation of efficient high-temperature fuel cells. The specific output power of fuel cells is still low. It is several times lower than the specific output power of internal-combustion engines. However, advances in electrochemistry and structural modernization of fuel cells will make the use of fuel cells for motor-vehicle transport and power engineering quite possible. The fuel cells are noiseless; they do not pollute the atmosphere.

In the USA the project FutureGen and in China the project GreenGen for the construction of power plants based on fuel cells are realized. Because of the project

GreenGen realization, total capacity of power plants in China will amount to 650 MW.

Focusing on the achievements of research and development, we can predict the further decrease in unit cost of alkaline fuel cells due to the development of large-scale production. These cells with capacity from 10 to 40 kW are already used in stationary and mobile power plants.

Because the FC permits not only generation of electric power from fuel, but also production of hydrogen from water in regeneration, it has a relatively broad spectrum of functional capabilities. Promising and already implemented applications of the FC (power range from mille watts to megawatts) are:

1. Stationary applications:

    - production of electricity (at power plants),
    - emergency power sources,
    - autonomous power supply.

2. Transport:

    - electric vehicles,
    - sea transport,
    - railway transport and mine equipment,
    - auxiliary transport (forklift trucks, airport equipment, and so on).

3. On-board power supply:

    - aviation and space,
    - submarines and maritime means of transport.

4. Mobile devices:

    - portable electronics,
    - mobile phones,
    - chargers for the army.

When using fuel cells in power engineering, problem arises of their connection with a grid to transfer electric power to it. FC is used with a three-phase grid when their power is above 10 kW. With an increase in power, DC-converters with galvanic uncoupling of the output buses from the power sources are mainly employed. Inverter directly connected to the grid or load is also based on completely controllable switches of the corresponding power [2].

Powerful hybrid systems (10 MW or more) operate with output voltages up to 10 kV, based on integrated gate-commutated turn-off thyristors (IGCT) and insulated gate bipolar transistors (IGBT). Micro turbine has a generator producing voltage sent to a three-phase rectifier and then to the common DC-buses of the system. The output inverter of the system, based on IGCT, corresponding to an input voltage of 6 kV.

Low-power systems (up to 10 kW) are usually employed with a single-phase grid. The source for the fuel cell is selected so as to ensure the mean rated power,

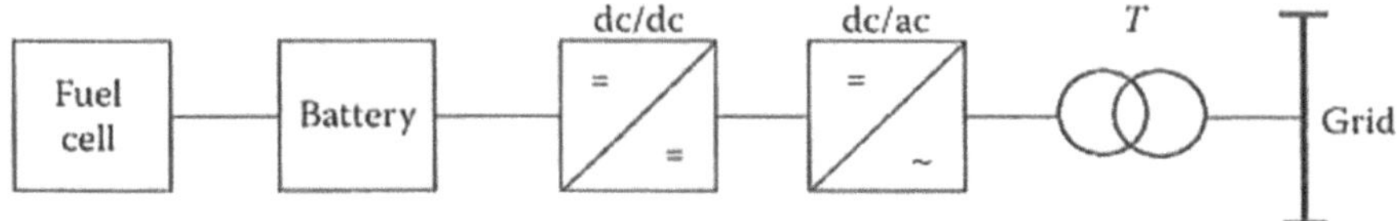

Fig. 9.7 Connection of the FC to the grid

whereas the peak loads must be covered by an energy store such as a battery. A simplified block diagram of the fuel cell connection is shown in Fig. 9.7.

The basic power components are a DC-converter and an inverter. To produce a more compact system, a high-frequency element is generally included. Its output is connected to a transformer T with the required output voltage. In such structures, both DC and AC-output channels are possible. To improve the size and mass of the unit, DC-converter with galvanic uncoupling may also be employed. In that case, an inverter without an output transformer is connected to the output of the DC-converter.

FC has been successfully used in uninterruptible power sources (UPSs). The FC is connected to a traditional UPS with battery storage. This significantly increases the possible operating period of the uninterruptible source.

Questions and Tasks to Chapter 9

1. What is understood by "unconventional methods of electric energy production"?
   (The physical and chemical effects are the basis of non-traditional methods of electricity production.)
2. Compare the total capacity of the existing non-conventional sources of electricity and traditional ones.
3. Explain the nature of the reaction of controlled thermonuclear fusion. What chemical elements can be used in this reaction? What is the principal difference between nuclear and thermonuclear reactors?
4. The principle of Tokamak working.
5. What parameters should have the deuterium-tritium plasma to initiate controlled nuclear fusion? What is it the Lowson criterion?
6. What physical effects lie in the basis of inertial (pulsed) thermonuclear fusion?
7. What results are achieved in the area of the compression fusion reactors development?
8. What are main obstacles to rapid putting into service these types of reactors?
9. Electrochemical energy sources: principles of operation, results achieved, and restrictions.
10. What is your opinion about the future of fuel cells?
11. Your opinion about the fields of the most effective fuel cells application in the electric power engineering.

References

1. Schwab AJ. Elektromagnetische Vertraglichkeit [in German], (Vierte neubearbeitete auflag). Berlin, Heidelberg: Springer; 1996.
2. Ben Elghali SE, Benbouzid MEH, Charpentier JF. Marine tidal current electric power generation technology: state of the art and current status. In: IEEE international conference on electric machines and drives 2007, IEMDC'07. vol. 2, p. 1407–12

Conclusion

In conclusion, I would like to formulate two points that, I hope, will help you to distinguish the reality from speculation that loaded mass media.

1. Currently, the main problem in the world energy is not a lack of energy resources, but lack of sufficient investment in their exploration and transformation into electrical and thermal energy. In the twenty-first century, the global shortage of energy resources does not threaten to humanity.

In the more distant future, output is likely to be found in the form of a transition to a fundamentally new technology, not yet known to us. The founder and head of Organization of Petroleum Exporting Countries (OPEC), Sheikh Ahmed Zaki Yamani, said: "The Stone Age ended not because the world ran out of stones. So the oil age will end not because oil is over, but because there will be new technologies."

2. A real threat to the sustainable development of civilization comes from increasing destructive impact of industry on the environment, primarily from companies and enterprises fuel and power complex.

In the textbook it has been shown on the basis of actual quantitative data that both problems are influencing just now on all sides of public life: politics, life standard, health of people, technology etc.

Present and, especially, next generations of scientists and experts have to be ready to eliminate these threats, to undertake measures for their removal or, at least, for their decrease. There are not so many methods (ways) to do this:

1. Increase of the efficiency of non-renewable power generating resources. It means energy—sawing behavior of the energy producers (suppliers) and energy consumers, which includes:

- application of energy saving technologies of energy production, transmission, distribution, and consumption;
- energy saving behavior of people in everyday life, including corresponding correction of the way of life.

V.Y. Ushakov, *Electrical Power Engineering*, Green Energy and Technology, https://doi.org/10.1007/978-3-319-62301-6

2. Radical enlargement of the scale of unconventional renewable energy sources use.
3. Search for and mastering of new (unconventional) methods of electric and thermal energy production.
4. The governments of different countries and the management of corresponding companies should render comprehensive support to the use of renewable energy sources and new ways of producing electricity, at least until they become sufficiently competitive.
5. Realization of measures for environmental protection from any kind of pollution.

To achieve the goals, they need to be fully realized.According to forecasts of scientists, the implementation of these measures will be the essence of *the global energy revolution,* which will happen in the twenty-first century and will be an essential part of the third industrial revolution.

Four groups of factors will determine the beginning and the pace of development:

(1) economic—the need for investment in large volumes,
(2) technological—requiring rapid change of technological modes,
(3) social—(a) rapid population growth in developing countries, and (b) inability to quickly change the existing way of life,
(4) global (in the historical and geographical aspects)—the uneven distribution of energy resources across countries and regions.

The most likely scenario for the development of civilization, in which electric power is mainly based on conventional primary energy—coal, gas, oil, and hydro and nuclear fuel, expanded use of renewable energy sources, and some of the known non-traditional ways to produce electricity. In the coming decades, any new sources of energy or fundamentally new methods of producing electricity and heat cannot be seen.

In the face of increasing globalization, the main feature of the electric power industry of the XXIst century will be further development of electric power systems and their integration as a cross-country in Eurasia and North America, as well as full implementation of the concept of Smart Grid.

To eliminate the global threat to peace, it is vital to create of a civilized energy market—the market—providing a balance of interests of buyers and sellers.